LEED

GREEN ASSOCIATE EXAM GUIDE (LEED GA)

Comprehensive Study Materials, Sample Questions, Mock Exam,
Green Building LEED Certification, and Sustainability

PLEASE
AS A COURTESY TO READERS AFTER YOU, DO NOT WRITE IN BOOK.
THANK YOU

GANG CHEN

ArchiteG®, Inc.
Irvine, California

LEED Green Associate Exam Guide (LEED GA):
Comprehensive Study Materials, Sample Questions, Mock Exam, Green Building LEED Certification, and Sustainability

v3.4   Issued on 1/23/2014

Copy Editor: Barry Wenger

ArchiteG®, Inc.
http://www.ArchiteG.com
http://www.GreenExamEducation.com

ISBN: 978-0-9843741-3-7

PRINTED IN THE UNITED STATES OF AMERICA

What others are saying about *LEED Green Associate Exam Guide…*

"Finally! A comprehensive study tool for LEED GA Prep!

"I took the 1-day Green LEED Green Associate course and walked away with a power point binder printed in very small print—which was missing MUCH of the required information (although I didn't know it at the time). I studied my little heart out and took the test, only to fail it by 1 point. Turns out I did NOT study all the material I needed to in order to pass the test. I found this book, read it, marked it up, retook the test, and passed it with a 95%. Look, we all know the LEED Green Associate exam is new and the resources for study are VERY limited. This one's the VERY best out there right now. I highly recommend it."
—**ConsultantVA**

"Complete overview for the LEED Green Associate exam

"I studied this book for about 3 days and passed the exam … if you are truly interested in learning about the LEED system and green building design, this is a great place to start."
—**K.A. Evans**

"I just finished taking the LEED Green Associate exam and, thankfully, I passed it on the first try by using this book as my primary study guide...I particularly liked the way the author organized the information within it. "
—**Lewis Colon**

"My name is Elizabeth (last name deleted to protect her privacy) and I am a junior at the University of....(college name deleted to protect her privacy). This summer I attained an internship with Kath Williams and Associates, a collaborative of creative independent contractors who come together to support innovative green projects, to learn about sustainable building. At the beginning of July I spoke with Kath about taking the LEED Green Associate Exam. After having no prior experience in sustainable architecture or LEED buildings I used both your review book and the study guides by USGBC as my study references. I wanted to e-mail to thank you for such a comprehensive review guide that was enormously helpful in passing the exam. Without your guide I don't know if I would have passed. Thank you so much. "
—**Elizabeth**

"A Great Book for Preparing the LEED Exam!

"I have read almost all the books for LEED Exams, and found LEED Exam Guide Series to be the best. The USGBC Reference Guide was too detailed and kind of confusing. Some other third party books have too many grammatical mistakes and are hard to understand, and way too many questions. The questions in those books are confusing instead of helpful. The USGBC workshop missed some of the very important information, like extra credits.
LEED Exam Guide Series gives you just the right amount of information for you to pass the LEED exam. Each book in the series includes study materials, sample questions and answers, as well as mock exam and answers for a specific LEED exam. It also gives you the most information for you to get your building LEED certified. A Great book! "
—**Ellen**

"A Wonderful Guide for the LEED Green Associate Exam

"After deciding to take the LEED Green Associate exam, I started to look for the best possible study materials and resources. From what I thought would be a relatively easy task, it turned into a tedious endeavor. I realized that there are vast amounts of third-party guides and handbooks. Since the official sites offer little to no help, it became clear to me that my best chance to succeed and pass this exam would be to find the most comprehensive study guide that would not only teach me the topics, but would also give me a great background and understanding of what LEED actually is. Once I stumbled upon Mr. Chen's book, all my needs were answered. This is a great study guide that will give the reader the most complete view of the LEED exam and all that it entails.

"The book is written in an easy-to-understand language and brings up great examples, tying the material to the real world. The information is presented in a coherent and logical way, which optimizes the learning process and does not go into details that will not be needed for the LEED Green Associate Exam, as many other guides do. This book stays dead on topic and keeps the reader interested in the material.

"I highly recommend this book to anyone that is considering the LEED Green Associate Exam. I learned a great deal from this guide, and I am feeling very confident about my chances for passing my upcoming exam."
—**Pavel Geystrin**

"Like other books in the LEED Exam Guide Series, this is a great timesaver! The important information that you need to memorize is already highlighted / underlined by the author. This can really save me a lot of time. I love it! A Great Timesaver! "
—**Alice**

"Easy to read, easy to understand

"I have read through the book once and found it to be the perfect study guide for me. The author does a great job of helping you get into the right frame of mind for the content of the exam. I had started by studying the Green Building Design and Construction reference guide for LEED projects produced by the USGBC. That was the wrong approach, simply too much information with very little retention. At 636 pages in textbook format, it would have been a daunting task to get through it. Gang Chen breaks down the points, helping to minimize the amount of information but maximizing the content I was able to absorb. I plan on going through the book a few more times, and I now believe I have the right information to pass the LEED Green Associate Exam."
—**Brian Hochstein**

"All in one—LEED GA prep material

"Since the LEED Green Associate exam is a newer addition by USGBC, there is not much information regarding study material for this exam. When I started looking around for material, I got really confused about what material I should buy. This LEED GA guide by Gang Chen is an answer to all my worries! It is a very precise book with lots of information, like how to approach the exam, what to study and what to skip, links to online material, and tips and tricks for passing the exam. It is like the 'one stop shop' for the LEED Green Associate Exam. I think this book can also be a good reference guide for green building professionals. A must-have!"
—**SwatiD**

Leadership in Energy and Environmental Design (LEED)

LEED-CERTIFIED   LEED-SILVER
LEED-GOLD   LEED-PLATINUM

LEED GREEN ASSOCIATE

LEED AP BD+C   LEED AP ID+C
LEED AP O+M
LEED AP HOMES   LEED AP ND

LEED FELLOW

Dedication

To my parents, Zhuixian and Yugen,
my wife Xiaojie, and my daughters
Alice, Angela, Amy, and Athena.

Disclaimer

This book provides general information about the LEED Green Associate Exam and green building LEED Certification. It is sold with the understanding that the publisher and author are not providing legal, accounting, or other professional services. If legal, accounting, and other professional services are required, seek the services of a competent professional firm.

It is not the purpose of this book to reprint the content of all other available texts on the subject. You are urged to read other available texts and tailor them to fit your needs.

Great effort has been made to make this book as complete and accurate as possible; however, nobody is perfect, and there may be typographical or other mistakes. You should use this book as a general guide and not as the ultimate source on this subject.

This book is intended to provide general, entertaining, informative, educational, and enlightening content. Neither the publisher nor the author shall be liable to anyone or any entity for any loss or damages, or alleged loss and damages, caused directly or indirectly by the content of this book.

USGBC and LEED are trademarks of the US Green Building Council. The US Green Building Council is not affiliated with the publication of this book.

If you do not wish to be bound by the above, you may return this book to the publisher for a full refund.

Legal Notice

All our books, including "ARE Mock Exams Series" and "LEED Exam Guides Series," are available at **GreenExamEducation.com**

Check out FREE tips and info at **GeeForums.com**, you can post your questions or vignettes for other users' review and responses.

Contents

18. How does LEED fit into the green building market?
19. What are the benefits of LEED certification for your building?
20. What is the procedure of LEED certification for your building?
21. How much is the building registration fee and how much is the building LEED certification fee?
22. How are LEED credits allocated and weighted?
23. Are there LEED certified products?

1. What is the scope of the LEED Green Associate Exam?
2. What is the latest version of LEED and when was it published?
3. How many possible points does LEED v3.0 have?
4. How many different levels of building certification does USGBC have?
5. What is the process for LEED certification? What are the basic steps for LEED certification?
6. What does the registration form include?
7. What is precertification?
8. What is CIR?
9. When do you submit a CIR?
10. What are the steps for CIR?
11. Will CIR guarantee a credit?
12. What tasks are handled by GBCI and what tasks are handled by USGBC?
13. What are MPRs?
14. What types of projects should use LEED-NC?
15. What types of projects should use LEED-CS?
16. What types of projects should use LEED for Schools?

1. What do green buildings address?
2. Key stake holders and an integrative approach
3. The mission of USGBC
4. The structure of LEED Rating System
5. LEED certification tools
6. **Specific Technical Information**

Note: Each credit category is described in a standard format, including Overall Purpose, Mnemonics, Core Concepts, Recognition, Regulation and Incentives, Overall strategies and technologies, and Specific Technical Information, and **Summary and Mnemonics**.

Specific Technical Information for **each credit** includes Purpose, Credit Path, Submittals, Synergies,

Possible Strategies and Technologies, Extra Credit (Exemplary Performance), Project Phase, <u>LEED Submittal</u> Phase, Related Code or Standard, and Responsible Party.

A few of the credits are for schools only or for CS only, and are <u>unlikely to be tested on the LEED Green Associate Exam</u>. We just list these credits' names and related points, omitting their detailed discussions.

Preface

Starting on December 1, 2011, GBCI will begin to draw LEED Green Associate Exam questions from the second edition of *Green Building and LEED Core Concepts Guide*. We have incorporated this latest information in our book, and changed our book title to *LEED Green Associate Exam Guide*. See Appendix 5 for Important Items Covered by the Second Edition of *Green Building and LEED Core Concepts Guide*.

There are two main purposes for this book: to help you pass the LEED Green Associate Exam and to assist you with understanding the process of getting your building LEED certified.

The LEED Green Associate Exam is the most important LEED exam for two reasons:

1. You have to pass it in order to get the title of LEED Green Associate.

2. It is also the required <u>Part One</u> (2 hours) of <u>ALL</u> LEED AP+ exams. You have to pass it plus Part Two (2 hours) of the specific LEED AP+ exam of your choice to get any LEED AP+ title unless you have passed the old LEED AP exam before June 30, 2009.

There are a few ways to prepare for the LEED Green Associate Exam:

1. You can take USGBC courses or workshops. You should take USGBC classes at both the 100 (Awareness) and 200 (LEED Core Concepts and Strategies) level to successfully prepare for the exam A one-day course can cost $445 if you register early enough, and can be as expensive as $495 if you miss the early bird special. You will also have to wait until the USGBC workshops or courses are offered in a city near you.

   OR
2. Take USGBC online courses. You can go to the USGBC or GBCI websites for information. The USGBC online courses are less personal and still expensive.

   OR
3. Read related books. Unfortunately, there is NO official USGBC book on the LEED Green Associate Exam. However, there are a few third party books on the LEED Green Associate Exam. *LEED Green Associate Exam Guide (LEED GA)* is one of the first books covering this subject and will fill in this blank to assist you with passing the exam.

To stay at the forefront of the LEED and green building movement and make my books more valuable to their readers, I sign up for USGBC courses and workshops myself, and I review the USGBC and GBCI websites and many other sources to get as much information as possible on LEED. *LEED Green Associate Exam Guide (LEED GA)* is a result of this very comprehensive research. I have done the hard work so that you can save time preparing for the exam by reading my book.

Strategy 101 for the LEED Green Associate Exam is that you must recognize that you have only a limited amount of time to prepare for the exam. So, you must concentrate your time and effort on the most important content of the LEED Green Associate Exam. To assist you with achieving this goal, the book is broken into two major sections: (1) the study materials and (2) the sample questions and mock exam.

Chapter One covers LEED Exam Preparation Strategies, Methods, Tips, Suggestions, Mnemonics and Exam Tactics to Improve Your Exam Performance.

Chapters Two and Three cover general information. I use the question and answer format to try to give you the most comprehensive coverage on the subject of the LEED AP exam. I have given you only the correct answers and information to save you time, i.e., you do not need to waste your time reading and remembering the wrong information. As long as you understand and remember the correct information, you can pass the test, no matter how the USGBC changes the format of the exam.

Chapter Four contains the LEED Green Associate Exam Technical Review, including Overall Purpose, Mnemonics, Core Concepts, Recognition, Regulation and Incentives, Overall Strategies and Technologies, and **Specific Technical Information**.

Specific Technical Information for **each credit** includes Purpose, Credit Path, Submittals, Synergies, Possible Strategies and Technologies, Extra Credit (Exemplary Performance), Project Phase, LEED Submittal Phase, Related Code or Standard, and Responsible Party.

A few of the credits are only for schools or CS, and are unlikely to be tested on the LEED Green Associate Exam. We just list these credits' names and related points, omitting their detailed discussions.

In the back section, you will find sample questions and a mock exam. These are intended to match the latest real LEED Green Associate Exam as closely as possible and assist you in becoming familiar with the format of the exam.

Most people already have some knowledge of LEED. I suggest that you use a highlighter when you read this book; you can highlight the content that you are not familiar with when you read the book for the first time. You can try to cover the answer and read a question first. If you can come up with the correct answer before you read the book, you do not need to highlight the question and answer. If you cannot come up with the correct answer before you read the book, then highlight that question. This way, when you do the review later and read the book for the second time, you can just focus on the portions that you are not familiar with and save yourself a lot of time. You can repeat this process with different colored highlighters until you are very familiar with the content of this book. Then, you will be ready to take the LEED Green Associate Exam.

The key to passing the LEED Green Associate Exam, or any other exam, is to know the scope of the exam, and not to read too many books. Select one or two really good books and focus on them. Actually understand the content and memorize it. For your convenience, I have underlined the fundamental information that I think is very important. You definitely need to memorize all the information that I have underlined. You should try to understand the content first, and then memorize the content of the book by reading it multiple times. This is a much better way than "mechanical" memory without understanding.

There is a part of the LEED Green Associate Exam that you can control by reading study materials: The section regarding the number of points and credit process for the LEED building rating system. You should become very familiar with every major credit category. You should try to answer all questions related to this part correctly.

There is also a part of the exam that you may not be able to control. You may not have done actual LEED building certification, so there will be some questions that may require you to guess. This is the hardest part of the exam, but these questions should be only a small percentage of the test if you are well prepared. You should eliminate the obvious wrong answers and then attempt an educated guess. There is no penalty for guessing. If you have no idea what the correct answer is and cannot eliminate any obvious wrong answer, then do not waste too much time on the question, just pick a guess answer. The key is, try to use the same guess answer for all of the questions that you have no ideas at all. For example, if you choose "a" as the guess answer, then you should be consistent and use "a" as the guess answer for all the questions that you have no ideas at all. That way, you likely have a better chance at guessing more correct answers.

The actual LEED Green Associate Exam has 100 multiple-choice questions and you must finish it within 2 hours. The raw exam score is converted to a scaled score ranging from 125 to 200. The passing score is 170 or higher. You need to answer about 60 questions correctly to pass. There is an optional 10-minute tutorial for computer testing before the exam and an optional 10-minute exit survey.

This is not an easy exam, but you should be able to pass it if you prepare well. If you set your goal for a high score and study hard, you will have a better chance of passing. If you set your goal for the minimum passing score of 170, you will probably end up scoring 169 and fail, and you will have to retake the exam again. That will be the last thing you want. Give yourself plenty of time and do not wait until the last minute to begin preparing for the exam. I have met people who have spent 40 hours preparing and passed the exam, but I suggest that you give yourself at least two to three weeks of preparation time. On the night before the exam, you should look through the questions on the mock exam that you did not answer correctly and remember what the correct answers are. Read this book carefully, prepare well, relax and put yourself in the best physical, mental and psychological state on the day of the exam, and you will pass.

Chapter 1
LEED Exam Preparation Strategies, Methods, Tips, Suggestions, Mnemonics and Exam Tactics to Improve Your Exam Performance

1. The nature of LEED exams and exam strategies
LEED Exams are standardized tests. They should be consistent and legally defensible tests.

The earliest standardized tests were the Imperial Examinations in China that started in 587. In Europe, traditional school exams were oral exam, and the first European written exams were held at Cambridge University, England in 1792.

Most exams test knowledge, skills, and aptitudes. There are several main categories of exams:
1) Math exams testing your abilities to do various calculations
2) Analytical exams testing skills in separating a whole (intellectual or substantial) into its elemental parts or basic principles
3) Knowledge exams testing your expertise, skills and understanding of a subject
4) Creativity exams testing your skills to generate new ideas or concepts
5) Performance exams such as driving tests or singing competitions

LEED exams test your knowledge of LEED as well as some very basic analytical and calculation skills (on average 1% to 2% of the exam questions require calculations).The LEED Green Associate Exam is like a history or political science test, and requires a lot of memorization of LEED information and knowledge.

All LEED exams test candidates' abilities at three hierarchical cognitive levels:
Recognition: ability to recall facts.
Application: ability to use familiar principles or procedures to solve a new problem.
Analysis: abilities to break the problem down into its parts, to evaluate their interactions or relationship, and to create a solution.

A LEED exam has 100 multiple-choice questions. You need to pick <u>one, two, three, or even four correct answers</u> (some questions have five choices), depending on the specific question.

The exam writers usually use errors that people are likely to make to create the incorrect choices (or **distracters**) to confuse exam takers, so you **<u>HAVE to read the question very carefully</u>**, pay special attention to words like <u>may</u>, <u>might</u>, <u>could</u>, etc. Creating effective <u>distracters</u> is a key to creating a good exam. This means that the more confusing a question is, the easier it is for the GBCI to separate candidates with strong LEED knowledge from the ones with weak LEED knowledge.

Since most LEED exam questions have four choices, the **distracters** in a strong LEED exam should be able to attract at least 25% of the weakest candidates to reduce the effectiveness of guessing.

For exam takers, it is to your advantage to guess for questions that you do not know, or are not sure about, because the exam writers expect you to guess. They are trying to mislead you with the questions to make your guessing less effective. If you do not guess and do not answer the questions, you will be at a disadvantage when compared with other candidates. Eliminate the obvious wrong answers and then try an educational guess. It is better to guess with the same letter answer for all the questions that you do not know, unless it is an obvious wrong answer and has been eliminated.

2. LEED exam preparation requires short-term memory

Now that you know the nature of the LEED Exam, you should understand that LEED Exam Preparation requires **Short-Term Memory**. You should schedule your time accordingly: in the early stages of your LEED exam Preparation, you should focus on understanding and an **initial** review of the material; in the late stages of your exam preparation, you should focus on memorizing the material as a **final** review.

3. LEED exam preparation strategies and scheduling

You should spend about 60% of your effort on the most important and fundamental LEED material, about 30% of your effort on sample exams, and the remaining 10% on improving your weakest areas, i.e., reading and reviewing the questions that you answered incorrectly, reinforcing the portions that you have a hard time memorizing, etc.

Do NOT spend too much time looking for obscure LEED information because the GBCI will HAVE to test you on the most **common** LEED information. At least 80% to 90% of the LEED exam content will have to be the most common, important and fundamental LEED knowledge. The exam writers can word their questions to be tricky or confusing, but they have to limit themselves to the important content; otherwise, their tests will NOT be legally defensible. At most, 10% of their test content can be obscure information. You only need to answer about 60% of all the questions correctly. So, if you master the common LEED knowledge (applicable to 90% of the questions) and use the guess technique for the remaining 10% of the questions on the obscure LEED content, you will do well and pass the exam.

On the other hand, if you focus on the obscure LEED knowledge, you may answer the entire 10% obscure portion of the exam correctly, but only answer half of the remaining 90% of the common LEED knowledge questions correctly, and you will fail the exam. That is why we have seen many smart people who can answer very difficult LEED questions correctly because they are able to look them up and do quality research. However, they often end up failing the LEED exam because they cannot memorize the common LEED knowledge needed on the day of the exam. The LEED exam is NOT an open-book exam, and you cannot look up information during the exam.

The **process of memorization** is like **filling a cup with a hole at the bottom**: You need to fill it faster than the water leaks out at the bottom, and you need to constantly fill it; otherwise, it will quickly be empty.

Once you memorize something, your brain has already started the process of forgetting it. It is natural. That is how we have enough space left in our brain to remember the really important things.

It is tough to fight against your brain's natural tendency to forget things. Acknowledging this truth and the fact that you cannot memorize everything you read, you need to focus your limited time, energy and brain power on the most important issues.

The biggest danger for most people is that they memorize the information in the early stages of their LEED exam preparation, but forget it before or on the day of the exam and still THINK they remember them.

Most people fail the exam NOT because they cannot answer the few "advanced" questions on the exam, but because they have read the information but can<u>NOT</u> recall it on the day of the exam. They spend too much time preparing for the exam, drag the preparation process on too long, seek too much information, go to too many websites, do too many practice questions and too many mock exams (one or two sets of mock exams can be good for you), and **spread themselves too thin**. They end up **missing out on the most important information** of the LEED exam, and they will fail.

The LEED Exam Guide Series along with the tips and methodology in each of the books will help you MEMORIZE the most important aspects of the test to pass the exam ON THE FIRST TRY.

So, if you have a lot of time to prepare for the LEED exam, you should plan your effort accordingly. You want your LEED knowledge to peak at the time of the exam, not before or after.

For example, <u>if you have two months to prepare for the exam</u>, you may want to spend the first month focused on <u>reading and understanding</u> all of the study materials you can find as your **initial** review. Also during this first month, you can start <u>memorizing</u> after you understand the materials as long as you know you HAVE to review the materials again later to <u>retain</u> them. If you have memorized something once, it is easier to memorize it again later.

Next, you can spend two weeks focused on <u>memorizing</u> the material. You need to review the material at least three times. You can then spend one week on <u>mock exams</u>. The last week before the exam, focus on retaining your knowledge and reinforcing your weakest areas. Read the mistakes that you have made and think about how to avoid them during the real exam. Set aside a mock exam that you <u>have not taken</u> and take it seven days before test day. This will alert you to your weaknesses and provide direction for the remainder of your studies.

<u>If you have one week to prepare for the exam</u>, you can spend two days reading and understanding the study material, two days repeating and memorizing the material, two days on mock exams, and one day retaining the knowledge and enforcing your weakest areas.

The last one to two weeks before the LEED exam is <u>absolutely</u> critical. You need to have the "do or die" mentality and be ready to study hard to pass the exam on your first try. That is how some people are able to pass the LEED exam with only one week of preparation.

4. **Timing of review: the 3016 rule; memorization methods, tips, suggestions, and mnemonics**
   Another important strategy is to review the material in a timely manner. Some people say that the best time to <u>review</u> material is between <u>30 minutes and 16 hours</u> (the **3016** rule) after you read it for the first time. So, if you review the material right after you read it for the first time, the review may not be helpful.

I have personally found this method extremely beneficial. The best way for me to memorize study materials is to review what I learn during the day again in the evening. This, of course, happens to fall within the timing range mentioned above.

Now that you know the **3016** rule, you may want to schedule your review accordingly. For example, you may want to read <u>new</u> study materials in the morning and afternoon, then after dinner do an <u>initial review</u> of what you learned during the day.

OR

If you are working full time, you can read <u>new</u> study materials in the evening or at night and then get up early the next morning to spend one or two hours on an <u>initial review</u> of what you learned the night before.

The <u>initial</u> review and memorization will make your <u>final</u> review and memorization much easier.

Mnemonics is a very good way for you to memorize facts and data that are otherwise very hard to memorize. It is often <u>arbitrary</u> or <u>illogical</u> but it works.

A good mnemonic can help you remember something for a long time or even a lifetime after reading it just once. Without the mnemonics, you may read the same thing many times and still not be able to memorize it.

There are a few common Mnemonics:
1) **<u>Visual</u>** Mnemonics: Link what you want to memorize to a visual image.
2) **<u>Spatial</u>** Mnemonics: link what you want to memorize to a space, and the order of things in it.
3) **<u>Group</u>** Mnemonics: <u>Break up</u> a difficult piece <u>into</u> several smaller and more <u>manageable groups or sets</u>, and memorize the sets and their order. One example is the grouping of the 10 digit phone number into three groups in the U.S. This makes the number much easier to memorize.
4) **<u>Architectural</u>** Mnemonics: A combination of <u>Visual</u> Mnemonics and <u>Spatial</u> Mnemonics and <u>Group</u> Mnemonics.

Imagine you are walking through a building several times, along the same path. You should be able to remember the order of each room. You can then break up the information that you want to remember and link them to several images, and then imagine you hang the images on walls of various rooms. You should be able to easily recall each group in an orderly manner by imagining you are walking through the building again on the same path, and looking at the images hanging on walls of each room. When you look at the images on the wall, you can easily recall the related information.

You can use your home, office or another building that you are <u>familiar with</u> to build an <u>Architectural</u> Mnemonics to help you to organize the things you need to memorize.

5) **<u>Association</u>** Mnemonics: You can <u>associate</u> what you want to memorize <u>with a sentence</u>, a similarly pronounced word, or a place you are familiar with, etc.
6) **<u>Emotion</u>** Mnemonics: Use emotion to fix an image in your memory.
7) **<u>First Letter</u>** Mnemonics: You can use the <u>first letter</u> of what you want to memorize <u>to construct a sentence or acronym</u>. For example, **"Roy G. Biv"** can be use to memorize the order of the 7 colors of the rainbow, it is composed of the first letter of each primary color.

You can use **<u>Association</u>** Mnemonics and memorize them as <u>all</u> the plumbing fixtures for a typical home, PLUS Urinal.

OR
You can use "Water S K U L" (**First Letter** Mnemonics selected from website below) to memorize them:

<u>Water </u>Closets
<u>S</u>hower
<u>K</u>itchen Sinks
<u>U</u>rinal
<u>L</u>avatory

Here is another example of **First Letter** Mnemonics:

Materials and Resources (MR) - Mandatory materials to be collected for recycling:

<u>P</u>eople <u>C</u>an <u>M</u>ake <u>G</u>reen <u>P</u>romises

<u>P</u>aper
<u>C</u>ardboard
<u>M</u>etal
<u>G</u>lass
<u>P</u>lastics

5. **The importance of good and effective study methods**
   There is a saying: Give a man a fish and you feed him for a day. Teach a man to fish and you feed him for a lifetime. I think there is some truth to this. Similarly, it is better to teach someone HOW to study than just give him good study materials. In this book, I give you good study materials to save you time, but more importantly, I want to teach you effective study methods so that you can not only study and pass LEED exams, but also so that you will benefit throughout the rest of your life for anything else you need to study or achieve. For example, I give you samples of mnemonics, but I also teach you the more important thing: HOW to make mnemonics.

   Often in the same class, all the students study almost the SAME materials, but there are some students that always manage to stay at the top of the class and get good grades on exams. Why? One very important factor is they have good study methods.

   Hard work is important, but it needs to be combined with effective study methods. I think people need to work hard AND work SMART to be successful at their work, career, or anything else they are pursuing.

6. **The importance of repetition: read this book <u>at least</u> three times**
   Repetition is one of the most important tips for learning. That is why I have listed it under a separate title. For example, you should treat this book as the core study materials for your LEED exam and you need to read this book <u>at least three times</u> to get all of its benefits:

   1) The first time you read it, it is new information. You should focus on understanding and digesting the materials, and also do an <u>initial</u> review with the **3016** rule.
   2) The second time you read it, focus on reading the parts <u>I</u> have already highlighted AND <u>you</u> have <u>highlighted</u> (the important parts and the weakest parts for you).

3) The third time, focus on <u>memorizing</u> the information.

Remember the analogy of the <u>memorization process</u> as **filling a cup with a hole on the bottom**? Do NOT stop reading this book until you pass the real exam.

7. When should you start to do sample tests and mock exams?

After reading the study materials in this book at least three times, you can start to do mock exams.

8. How much time do you need for LEED exam preparation?

Do not give yourself too much time to prepare for the LEED exam. Two months is probably the maximum time you should allow for preparing for the LEED exam.

Do not give yourself too little time to prepare for LEED exam. You want to force yourself to focus on the LEED exam but you do NOT want to give yourself too little time and fail the exam. One week is probably the minimum time you should allow for preparing for the LEED exam.

9. The importance of a routine

A routine is very important for studying. You should try to set up a routine that works for you. First, look at how much time you have to prepare for the LEED exam, and then adjust your current routine to include LEED exam preparation. Once you set up the routine, stick with it.

For example, you can spend from 8:00 a.m. to 12:00 noon, and 1:00 p.m. to 5:00 p.m. on studying new LEED exam study materials, and 7:00 p.m. to 10:00 p.m. to do an initial review of what you learned during the daytime. Then, switch your study content to mock exams, memorization and retention when it gets close to the exam date. This way, you have 11 hours for LEED exam preparation everyday. You can probably pass the LEED exam in one week with this method. Just keep repeating it as a way to <u>retain</u> the LEED knowledge.

OR

You can spend 7:00 p.m. to 10:00 p.m. on studying new LEED exam study materials, and 6:00 a.m. to 7:00 a.m. to do an initial review of what you learned the evening before. This way, you have four hours for LEED exam preparation every day. You can probably pass LEED in five weeks with this LEED preparation schedule.

A routine can help you to memorize important information because it makes it easier for you to concentrate and work with your body clock.

Do NOT become panicked and change your routine as the exam date gets closer. It will not help to change your routine and pull all-nighters right before the exam. In fact, if you pull an all-nighter the night before the exam, you may do much worse than you would have done if you kept your routine. All-nighters or staying up late are not effective. For example, if you break your routine and stay up one-hour late, you will feel tired the next day. You may even have to sleep a few more hours the next day, adversely affecting your study regimen.

10. The importance of short, frequent breaks and physical exercise

Short, frequent breaks and physical exercise are VERY important for you, especially when you are spending a lot of time studying. They help relax your body and mind, making it much easier for you to concentrate when you study. They make you more efficient.

Take a five-minute break, such as a walk, at least once every one to two hours. Do at least 30 minutes of physical exercise every day.

If you feel tired and cannot concentrate, stop, go outside, and take a five-minute walk. You will feel much better when you come back.

You need your body and brain to work well to be effective with your studying. Take good care of them. You need them to be well-maintained and in excellent condition. You need to be able to count on them when you need them.

If you do not feel like studying, maybe you can start a little bit on your studies. Just casually read a few pages. Very soon, your body and mind will warm up and you will get into study mode.

Find a room where you will NOT be disturbed when you study. A good study environment is essential for concentration.

11. A strong vision and a clear goal

You need to have a strong vision and a clear goal: to master the LEED system and become very familiar with the LEED certification process. This is your number one priority. You need to master the LEED knowledge BEFORE you do any sample questions or mock exams. It will make the process much easier. Otherwise, there is nothing in your brain to be tested. Everything we discuss in this chapter is to achieve this goal.

As I have mentioned on many occasions, and I say it one more time here because it is so important:

It is how much LEED information you can understand, digest, memorize, and firmly retain that matters, not how many books you read or how many sample tests you have taken. The books and sample tests will NOT help you if you cannot understand, digest, memorize, and retain the important information for the LEED exam.

Cherish your limited time and effort and focus on the most important information.

Chapter 2
Overview

1. **What is LEED? What is the difference between LEED, LEED AP, LEED Green Associate, LEED AP+ and LEED Fellow?**
   Answer: LEED is a term for <u>buildings</u>. It stands for the Leadership in Energy and Environmental Design (LEED) Green Building Rating System™. It is a voluntary system set up by the U.S. Green Building Council (USGBC) to measure the sustainability and performance of a building.

   LEED AP, LEED Green Associate, LEED AP+, and LEED Fellow are terms for <u>people</u>.

   LEED AP stands for LEED Accredited Professionals. A LEED AP is a person who has passed at least one of the three <u>old</u> versions of the LEED exams (<u>LEED-NC, LEED-CI, and LEED-EB</u>) before June 30, 2009 and has the skills and knowledge to encourage and support integrated design, to take part in the design process, and to control the application and certification process for a LEED building.

   LEED APs have **three choices:**

   1) <u>Do nothing</u> and keep the title of LEED AP. It is also called LEED AP <u>without</u> specialty, or a legacy LEED AP.

   2) Starting June 2009, a LEED AP can choose to <u>enroll in</u> the new tiered system, accept the GBCI <u>disciplinary policy</u>, and **complete** the prescriptive <u>Credentialing Maintenance Program</u> (**CMP**) for the initial two-year reporting period. After opting in, a LEED AP can use one of the new specialty designations (BD+C, ID+C, O+M) after his name. The LEED AP must opt in before summer of 2011.

   3) Starting June 2009, a LEED AP can choose to <u>opt in</u>, accept the GBCI <u>disciplinary policy</u>, **agree** to CMP, and **pass** Part Two of one of the LEED AP+ specialty exams to become a LEED AP+. The LEED AP only needs to take Part Two of the LEED AP+ exam if he takes the exam <u>by summer, 2011</u>. After opting in, he can use one of the new specialty designations (Building Design and Construction, or BD+C, Interior Design and Construction, or ID+C, Building Operation and Maintenance, or O+M) after his name.

   By choosing Paths 2 or 3 above, a LEED AP will become a LEED AP+ or a LEED AP with Specialty. See detailed information below:
   A LEED AP who passed the <u>old</u> LEED-<u>NC</u> exam will become a LEED AP <u>BD+C</u>, or a LEED AP <u>BD&C</u>.

   A LEED AP who passed the <u>old</u> LEED-<u>CI</u> exam will become a LEED AP <u>ID+C</u> or a LEED AP <u>ID&C</u>.

A LEED AP who passed the old LEED-EB exam will become a LEED AP O+M or a LEED AP O&M.

Maintenance: After completing one of these three choices, a LEED AP will be treated like any other LEED AP+ and will need to pay a $50 fee and take 30 hours of required class every two years to maintain the title of LEED AP+. The $50 fee is waived for the first two years for a Legacy LEED AP who decided to opt in.

LEED Green Associate, LEED AP+ and LEED Fellow are the three new tiers of professional credentials for LEED professionals. The GBCI started to use these new designations in 2009.

A **LEED Green Associate** is a green building professional with a basic level of LEED knowledge, i.e., a person who has passed the LEED Green Associate Exam and possesses the skills and knowledge to understand and support LEED projects and green building in the areas of design, construction, operation, and maintenance, AND has signed the paperwork to accept the GBCI disciplinary policy.

The LEED Green Associate Exam will NOT test the detailed information for each LEED credit, i.e., WE2.1, WE2.2, etc. It will test you on the overall core concepts. You need to know what the strategies are for the overall categories of WE, EA, etc.

However, it IS much easier to understand, digest and organize LEED information, and it will be to your advantage to learn and memorize the LEED information for each specific LEED credit. You will be a much more desirable and useful support member for a LEED project team. You will definitely be able to answer generic LEED questions for the major LEED categories and pass the LEED Green Associate Exam.

You also need to know the codes and regulations related to each of the main LEED categories, i.e., WE or EA, etc., but NOT information for each credit within the category.

Exam Cost: The cost of LEED Green Associate Exam is a $50 non-refundable application fee plus $150 for USGBC members or full-time students and $200 for non-members.

Maintenance: A LEED Green Associate will need to pay a $50 fee and take 15 hours of required class every two years to maintain the title.

LEED AP+ is a green building professional with an advanced level of LEED knowledge, i.e., a person who has passed the LEED Green Associate Exam (or Part One of the LEED AP+ exams) and Part Two of a LEED AP+ Specialty Exam based on one of the LEED rating systems or the equivalent, AND has signed paperwork to accept the GBCI disciplinary policy.

There are five different categories of LEED AP+ specialty exams (Part Two of LEED AP+ exams) and five categories of related LEED AP+ credentials, including:

    LEED AP ID+C (Interior Design and Construction) LEED AP Homes
    LEED AP O+M (Operation and Maintenance)
    LEED AP ND (Neighborhood Development)
    LEED AP BD+C (Building Design and Construction)

For example, if you want to become a LEED AP BD+C, you need to pass the LEED Green Associate Exam (Part <u>One</u> of ALL LEED AP+ Exams) and the LEED BD+C Specialty Exam (Part <u>Two</u> of LEED AP+ exam specializing in LEED BD+C).

Both Part One and Part Two of the LEED AP+ exam have <u>100</u> multiple-choice questions, asking for one, two, three or even <u>four</u> correct answers (Some questions have <u>five</u> choices).

Exam Cost: The cost of the LEED AP+ exam is a <u>$100</u> non-refundable application fee plus:
1) If you are taking the combined exam (both Part One and Part Two): <u>$300</u> for USGBC members and <u>$450</u> for non-members.
2) If you are taking only the Part Two specialty exam: <u>$150</u> for USGBC members and <u>$250</u> for non-members.

Maintenance: A LEED AP+ will need to pay a <u>$50</u> fee and take <u>30</u> hours of required class every <u>two</u> years to maintain the title.

A **LEED Fellow** is a green building professional with an <u>extraordinary</u> level of LEED knowledge, and has made major contributions to green building industry. The GBCI is still working on the development of criteria for becoming a LEED Fellow.

Note: These are current fees. They may be changed by the GBCI at a later point, check the GBCI website for exact fees.

2. **Why did the GBCI create the new three-tier LEED credential system?**
Answer: To pursue on-going improvement and excellence, to assure stakeholders of LEED professionals' current competence and latest knowledge in green building practice, and meet <u>three</u> prevailing market challenges: <u>stay current</u>, <u>differentiation</u> and <u>specialization</u>.

3. **Do I need to have LEED project experience to take the LEED exams?**
Answer: If you are taking LEED Green Associate Exam, you need to meet <u>one of the three</u> criteria below:

1) Studying in an education program that addresses green building principles

    Per USGBC:
    "GBCI has confirmed that all of USGBC's education programs, with the exception of its webinars, satisfy this requirement." Refer to the link below for more information:
    http://www.usgbc.org/DisplayPage.aspx?CMSPageID=2011

    This means ANYONE can take the LEED Green Associate Exam.

    Why?
    Because ANYONE can take the one of the USGBC's education programs listed at link above and qualify for the exam as long as he is willing to pay the fee (about a few hundred dollars, the exact fee depends on which program he chooses). A candidate probably will take one of these education programs to prepare for the LEED exam anyway.

OR
2) Working in a sustainable field

OR
3) Have previous experience supporting a LEED-registered project

Your above-mentioned experience needs to be confirmed by a dated letter in 750 words or less on a letterhead, signed by a client, supervisor, project manager, teacher, or someone else who is qualified to evaluate your performance. The letter author needs to use a business card or other means to demonstrate his title and relationship to the applicant.

The letter must also provide the date of the applicant's related experience, and confirm and summarize the applicant's participation or experience in one of the three qualifying categories (green building education, LEED project, or sustainable work).

For a candidate qualifying through green building education, an official transcript or a certificate of completion can substitute for a letter of attestation.

5% to 7% of the candidates will be audited for the above requirements.
You also need to agree to the GBCI credential maintenance requirements and the disciplinary policy.

If you are taking the LEED AP+ exam, you must have documented experience on one or more LEED projects within three years of your exam application submittal date.

4. **How do I become a LEED AP+? Do I have to take the LEED Green Associate Exam first to become a LEED AP+?**
   Answer: There are four paths to become a LEED AP+. See below. Paths One and Two do NOT require taking the LEED Green Associate Exam first, while Paths Three and Four do.

   Paths One and Two: If you are one of those lucky people who passed one of the three old LEED AP exams before June 30, 2009, and decided to opt into the new tiered system, you just need to opt in and accept the GBCI disciplinary policy and CMP requirements, and
   1) **pass** the Part Two of the LEED AP+ exam before summer of 2011 and **agree** to the CMP

   OR
   2) **complete** the CMP before summer of 2011.

   Path Three: If you have already passed the LEED Green Associate Exam, you only need to pass one of the specialty exams (Part Two of the LEED AP+ exams) to earn the title of one of the LEED AP with specialty.

   Path Four: If you have not taken any LEED exams, you need take both parts of the LEED AP+ exams. You need to take both parts at the same time (back to back in the same sitting), UNLESS you have passed one of sections before. In this case, you can schedule to retake the failed section only at a different time for an additional fee.

5. **How many questions do you need to answer correctly to pass the LEED exams?**
   Answer: Many readers have asked me this question before. The short answer is about 60 correct questions or 60% of the 100 total questions. The only official answer you can get is from the GBCI,

but they do NOT give out an exact number. So, different people have different opinions. Here is my justification:

1) The GBCI <u>intends</u> to use 60% of the correct questions as the benchmark for a passing score. Here is a simple calculation:

LEED Exams
Maximum Score: 200
Minimum Score: 125
Difference: 200-125 = 75 points (The score difference between someone who answers everything <u>correctly</u> and someone answers everything <u>incorrectly</u>.)

75 × 60% = 45 (points)
125 + 45 = 170 points = passing score

Total questions: 100
60% × 100 = 60 = correct answers needed to pass the LEED exams.

2) The <u>exact</u> number may be different depending on the difficulty of the version of the exam that you are taking.

Why?
Because GBCI wants the LEED passing score to be legally defensible. So, it uses subject matter experts to set the minimum level of required LEED knowledge and professional psychometricians to analyze the performance of people taking the beta tests, and uses the <u>Angoff Method</u> to decide the final passing score.

The <u>easier</u> versions of the test will need a <u>higher</u> number of correct answers to pass, and the <u>harder</u> versions of the test will need a <u>lower</u> number of correct answers to pass. So, the end result is the correct number of the questions needed to pass the exam is about 60, but probably NOT exactly 60.

If you reach the required level of LEED knowledge, no matter which version of the test you take, you should pass.

6. **What are the key areas that USGBC uses to measure the performance of a building's sustainability?**
Answer: Sustainable Site development (<u>S</u>S), Water Efficiency (<u>WE</u>), Energy and Atmosphere (<u>E</u>A), Materials and Resource (<u>M</u>R), Indoor Environmental Quality (<u>I</u>EQ), Innovations and Design process (<u>I</u>D), and Regional Priority (<u>R</u>P).

Mnemonics: <u>S</u>hall <u>WE</u> <u>M</u>ake <u>I</u>t <u>I</u>talic and <u>R</u>ed?

7. **How many LEED exams does USGBC have?**
Answer:
1) Before 2009, there were three <u>old</u> versions of the LEED exams: LEED New Construction (LEED-NC) v2.2, LEED Commercial Interior (LEED-CI) v2.0, and LEED Existing Building (LEED-EB) v2.0. You just needed to pass <u>one</u> of the three exams to earn the <u>old</u> title of LEED AP.
2) In 2009, GBCI started to introduce a <u>new</u> three-tier LEED credential system, and related <u>new</u> versions of LEED exams, including:

    a.   The LEED Green Associate Exam. It is also the Part One of <u>ALL</u> LEED AP+ exams.

    b.   Part Two of LEED AP+ exams, including the following specialties (exam takers just need to choose one):

        LEED AP ID+C (Interior Design and Construction)
        LEED AP Homes
        LEED AP O+M (Operation and Maintenance)
        LEED AP ND (Neighborhood Development)
        LEED AP BD+C (Building Design and Construction)

8. Are LEED Exams valid and reliable?
Answer: Yes. They are valid because they can measure what they intend to measure. They are reliable because they can accurately measure a candidate's skills.

9. How many member organizations does the USGBC have?
Answer: More than <u>11,000</u> member organizations.

10. How many regional chapters does the USGBC have?
Answer: <u>75</u>.

11. What is the main purpose of the USGBC?
Answer: To improve the way a building is designed, built and used to achieve a healthy, profitable and environmentally responsible building and environment, and to improve the quality of life.

12. What are the guiding principles of the USGBC?
Answer: The USGBC emphasizes not only the decisions themselves but also *how* the decisions are made. Its guiding principles are: advocate the **triple bottom lines** (balance <u>environmental</u>, <u>social</u> and <u>economic</u> needs; some people summarize these as <u>planet</u>, <u>people</u> and <u>profit</u>); build leadership; strive to achieve the balance between humanity and nature; uphold integrity and restore, preserve and protect the environment, species and ecosystem; use democratic and interdisciplinary approaches to ensure inclusiveness to achieve a common goal; openness, honesty and transparency.

13. How much energy and resources do buildings consume in the US?
Answer: Per <u>the </u>USGBC and US Department of Energy, buildings consume about <u>39%</u> of total energy, <u>74%</u> of electricity and <u>1/8</u> of the water in the US. Buildings also use valuable land that could otherwise provide ecological resources. In 2006, more than <u>1 billion metric tons</u> of carbon dioxide was generated by the commercial building sector. That is over a <u>30%</u> increase from the 1990 level.

14. What is the most important step to get your building certified?
Answer: It is to register your building with USGBC online at usgbc.org.

15. What are the benefits of green buildings and LEED Certification?
Answer: <u>To</u> enhance the building and company marketability and provide <u>branding opportunities</u>, positive impact on health and environment, may increase occupants' <u>productivity</u>, <u>reduce</u> building operating <u>costs</u>, and help to create <u>sustainable communities</u>.

16. **Who develops the LEED green building rating systems?**
    Answer: The USGBC committee and volunteers.

17. **What current reference guides and specific green building rating systems does the USGBC have?**
    Answer: The USGBC has the following reference guides and green rating system portfolio:

    1) *The LEED 2009 Reference Guide for Green Building Design and Construction (BD&C)*, which covers the following LEED rating systems:
       • LEED for New Construction (LEED-NC)
       • LEED for Core and Shell (LEED-CS);
       • LEED for Schools
       • LEED for Health Care*
       • LEED for Retails*

    2) *The LEED 2009 Reference Guide for Green Interior Design and Construction (ID&C)*, which covers the following LEED rating systems:
       • LEED for Commercial Interiors (LEED-CI)
       • LEED for Retails Interiors*

    3) *The LEED 2009 Reference Guide for Green Building Operations and Maintenance (O&M)*, which covers the following LEED rating systems:
       • LEED for Existing Buildings (LEED-EB)
       • LEED for Existing Schools*

       Note: These two LEED-EB rating systems are the only systems that cover building operation. All other LEED systems cover building design and construction, but NOT operation.

    4) *The LEED for Neighborhood Development Reference Guide (LEED-ND)*.

    5) **Application Guide for Multiple Buildings and On-Campus Building Projects (AGMBC)**, 2010 Edition. See link below:
       http://www.usgbc.org/campusguidance

    6) USGBC has **NOT** updated the following LEED rating systems or related information to LEED 2009 or LEED v3.0:
       • LEED for Homes (**LEED-H**): The new version will be released in 2011.
       • LEED for Laboratories*

       *These LEED rating systems are under development or currently in a pilot program. Once their new versions are finalized, the USGBC will sell supplements to their reference guides for them.

    The complete sample LEED 2009 credit templates will be available in the future at link below:
    http://www.usgbc.org/DisplayPage.aspx?CMSPageID=1447

    See link below for various LEED rating systems:
    http://www.usgbc.org/projecttools

After <u>June 26, 2009</u>, ALL buildings have to be registered under LEED 2009 or v3.0.

18. How does LEED fit into the green building market?
Answer: LEED fits into this market by providing voluntary, market-driven, and consensus-based rating systems. It is based on accepted environmental and energy principles, and maintains a balance between emerging concepts and established practices. Green buildings developed under the LEED rating systems can reduce operating costs and create branding and marketing opportunities for buildings and organizations. They are good for the environment and public health, increase occupant productivity and create sustainable communities.

19. What are the benefits of LEED certification for your building?
Answer:
a. The building is qualified for various government initiatives.
b. The building can obtain USGBC (third party) validation of your achievement.
c. You can be recognized for your commitment to environment issues.
d. <u>Branding opportunities</u> and market exposure through media, Greenbuild conference, USGBC, cases studies, etc.

20. What is the procedure of LEED certification for your building?
Answer:
a. Go to www.gbci.org to <u>register</u> your building, the earlier the better. This is the <u>most important</u> step.
b. You must document that your building meets prerequisites and a minimum number of points to be certified.
Refer to the LEED project checklist for the points needed for various levels of LEED certification.

21. How much is the building registration fee and how much is the building LEED certification fee?
Answer: The building <u>registration</u> fee is $450 for USGBC members and $600 for non-members.

The building <u>certification</u> fee varies, but it is $2000 on average.

22. How are LEED credits allocated and weighted?
Answer: They are allocated and weighted per strategies that will have greater positive impacts on the most important environmental factors: <u>CO_2 reductions</u> and <u>energy efficiency</u>. They are weighted against **13 aftereffects of human activities**, including <u>carbon footprint / climate change (25%),</u> <u>indoor-air quality (15%)</u>, resource/fossil-fuel depletion (9%), particulates (8%), water use/water intake (7%), human health: cancer (7%), ecotoxity (6%), eutrophication (5%), land use/habitat alteration (5%), human health: non cancer (4%), smog formation (4%), acidification (3%), and ozone depletion (2%) .

These <u>13</u> aftereffects are created by the U.S. <u>Environmental Protection Agency</u> (**EPA**), and are also known as "**TRACI**," a **mnemonic** for "<u>T</u>ool for the <u>R</u>eduction and <u>A</u>ssessment of <u>C</u>hemical and Other Environmental <u>I</u>mpacts."

1) The USGBC uses a reference building to estimate environmental impact in the 13 categories above.

2) The USGBC also used a tool developed by the <u>National Institute of Standard and Technology</u> (**NIST**) to <u>prioritize</u> the TRACI categories.

3) It also created a matrix to show the existing LEED credits and the TRACI categories, and used data that quantify building impacts on human health and environment to allocate points for each credit.

<u>The points for Energy</u> and <u>transportation</u> credits have been greatly increased in LEED 2009 primarily because of the importance of <u>reducing carbon or greenhouse gas emissions</u>. <u>Water Efficiency</u> is also a big winner in the credits, doubled from 5 to 10 points for some LEED rating systems.

In addition to the weighting exercise, the USGBC also used value judgments because if it only used the TRACI–NIST tool, some existing credits would be worth almost nothing, like the categories for <u>human health</u> and <u>indoor air quality</u>. The USGBC wanted to keep the LEED system somewhat consistent and retained the existing credits including those with few environmental benefits. So each credit was assigned <u>at least one point</u> in the new system.

There are NO negative values or fractions for LEED points.

<u>For instance, 20%</u> reduction of <u>indoor</u> water-use used to be able to gain points, now this is a prerequisite in LEED 2009.

23. Are there LEED certified products?
Answer: <u>ALL</u> LEED systems certify buildings or projects, <u>NOT</u> products. Products can <u>contribute</u> to a LEED project, and sometimes a product's data is required as part of a LEED submittal package.

Chapter 3
Introduction to
The LEED Green
Associate Exam

1. **What is the scope of the LEED Green Associate Exam?**
   Answer: Per the USGBC, the content of the LEED Green Associate Exam is limited to:

   a. **Understanding the <u>fundamental</u> credit intents, strategies, requirements, technologies, and submittals for the six <u>major</u> credit categories:** <u>Sustainable site development (SS), water savings/efficiency (WE), energy and atmosphere (EA), materials and resource (MR), indoor environmental quality (IEQ), and innovations and design process (ID), as well as innovation in upgrades, maintenance, and operations.</u>

   b. **<u>Process</u> of LEED application and opportunities for <u>synergy</u>**
   Synergy means the combined effects of two or more agents or forces greater than the sum of their individual.
   LEED Green Associate Exam may test the following aspects:
   - Site, budget, schedule, program, and other <u>requirements</u>
   - Hard, soft, and life-cycle <u>costs</u>
   - Environmental Building News, **USGBC**, **NRDC** (Natural Resources Defense Council), and other <u>green resources</u>
   - **Green Seal**, **SMACNA** (Sheet Metal and Air Conditioning Contractors National Association Guidelines), **ASHRAE** (American Society of Heating, Refrigeration and Air-conditioning Engineers), and other <u>standards</u> for LEED credit
   - Waste management, IEQ and energy, and other <u>interaction</u> between LEED Credits
   - **<u>CIR</u>** (Credit Interpretation Requests/Rulings) and previous <u>samples</u> leading to extra points
   - Project <u>registration</u> and **LEED Online**
   - <u>Score Card</u> for LEED
   - Supplementary documentation, project calculations, and other <u>Letter Templates</u>
   - LEED credit **strategies**
   - Property, project, and LEED <u>boundaries</u>
   - <u>Minimum</u> LEED certification program requirements and/or **prerequisites**
   - Certification <u>goal</u> and preliminary rating
   - Opportunities for the same building to get <u>multiple certifications</u>, i.e., commercial interior as well as core and shell
   - Certified building in LEED neighborhood development
   - Operations and maintenance for certified new building construction
   - Requirements for <u>occupancy</u> (e.g., an existing building <u>must be fully occupied for 12 continuous months</u> as described in minimum program requirements)
   - Logo usage, trademark usage, and other USGBC <u>policies</u>

- Requirements on receiving LEED AP <u>Credit</u>

c. **Site factors of a project:**
1) **Connectivity** for Communities
   a) Carts, shuttles, car-sharing membership (e.g. Zipcar™), bike storage, public transportation, fuel efficient vehicle parking, parking capacity, carpool parking, and other means to improve **transportation**
   b) Ramps, crosswalks, trails, and other means to improve **pedestrian access** and circulation

2) **Zoning requirements** (calculations of site area, floor to area ratio, and other density components, building footprint, open space, development footprint, construction limits and specific landscaping restrictions)

3) **Development**
   Heat Islands including albedo, emissivity, non-roof/roof heat island effect; green roofs, Solar Reflectance Index (SRI), etc.

d. **Management of water:**
1) Understand <u>Quality and Types</u> of Water, i.e., blackwater, graywater, storm water and potable water
2) Understand Water <u>Management</u>, i.e., use low-flush fixtures such as water closets, urinals, sinks, lavatory faucets and showers to reduce water use; calculations of FTE (Full Time Equivalent); <u>baseline water demand</u>; irrigation; rainwater harvesting

e. **Project system and related energy impact:**
1) Ozone depletion, chlorofluorocarbon (CFC) reduction, no refrigerant option, fire suppressions without halons or CFC's, phase-out plan, Hydrochlorofluorocarbons (HCFC) and other <u>Environmental Concerns</u>
2) Green-e providers, off-site generated, renewable energy certificates and other <u>Green Power</u>

f. **Project materials acquisition, installation and management:**
1) Commingled, pre-consumer, post-consumer, collection and other requirements regarding <u>recycled materials</u>
2) <u>Locally</u>/regionally manufactured and harvested materials
3) Accounted by weight or volume, written plan, polychlorinated biphenyl (PCB) removal and asbestos-containing materials (ACM) management, reduction strategies and other requirements regarding <u>construction waste management</u>

g. **Project team coordination, stakeholder involvement in innovation and regional design:**
1) Civil engineer, landscape architect, architect, heating-ventilation-air-conditioning (HVAC) engineer, contractor, facility manager and related <u>integrated project team criteria</u>
2) Building reuse, material lifecycle and other <u>durability, planning, and management</u>
3) Appropriate and established requirements, regional green design and construction and other <u>innovative and regional design</u> measures

h. **Codes and regulations, public outreach and project surroundings:** building, electrical, mechanical, plumbing, fire protection codes, etc.

i. **Ability to <u>support the coordination of team and project</u> and <u>assist</u>** with the process of gathering the necessary requirements and information for the **LEED process** and coordinating the different job functions for **LEED certification**

    j. **Ability to <u>support the process</u> of LEED <u>implementation</u>**

    k. **Ability to <u>support</u> technical <u>analyses</u> for LEED <u>credits</u>**

2. What is the latest version of LEED and when was it published?
    a. **When was LEED v1.0 <u>released</u>?**
        Answer: It was first <u>launched</u> in August 1998, but officially <u>released</u> in <u>1999</u>.

    b. **When was LEED-NC v2.0 released?**
        Answer: It was first <u>published</u> in <u>1999</u>, but officially <u>released</u> in <u>March 2000</u>.

    c. **When was LEED-NC v2.1 released?**
        Answer: <u>November 2002</u>.

    d. **When was LEED-NC v2.2 <u>published</u>?**
        Answer: It was first <u>published</u> in <u>2003</u>, but officially <u>released</u> in <u>2005</u>.

    e. **When was LEED v3.0 launched?**
        Answer: On <u>4/27/2009</u>, the USGBC launched **LEED v3.0**, including: a new building certification model, a new LEED Online, and LEED 2009 (Improvement to LEED rating systems).

        LEED v3.0 is a part of the continuous evolution of the LEED building rating system. For LEED v3.0, the USGBC is trying to synchronize the prerequisites and credits of the different versions of the LEED systems, and to create a predictable LEED development cycle (like the building codes, probably every three years), a transparent environmental/human impact credit weighting (redistributing the available points in LEED), as well as regionalization (regional bonus credits). See the Appendixes of this book for links and more information on LEED v3.0.

3. How many possible points does LEED v3.0 have?
    Answer: Starting with LEED v3.0 or LEED 2009, **ALL** updated rating systems have or will have **110** possible points, including **100** possible base points and **10** bonus points.

    For LEED 2009 for **New Construction and Major Renovations (NC)**, there are <u>26</u> points for Sustainable Sites (SS), <u>10</u> points for Water Efficiency (WE), <u>35</u> points for Energy and Atmosphere (EA), <u>14</u> points for Materials and Resource (MR), <u>15</u> points for Indoor Environmental Quality (IEQ), <u>6</u> bonus points for Innovations in Design (ID), and <u>4</u> bonus points for Regional Priority (RP). For LEED 2009 for **Core and Shell (CS),** there are <u>28</u> points for Sustainable Sites (SS), <u>10</u> points for Water Efficiency (WE), <u>37</u> points for Energy and Atmosphere (EA), <u>13</u> points for Materials and Resource (MR), <u>12</u> points for Indoor Environmental Quality (IEQ), <u>6</u> bonus points for Innovations in Design (ID), and <u>4</u> bonus points for Regional Priority (RP).

    For LEED 2009 for **Schools**, there are <u>24</u> points for Sustainable Sites (SS), <u>11</u> points for Water Efficiency (WE), <u>33</u> points for Energy and Atmosphere (EA), <u>13</u> points for Materials and Resource (MR), <u>19</u> points for Indoor Environmental Quality (IEQ), <u>6</u> bonus points for Innovations in Design

(ID), and <u>4</u> bonus points for Regional Priority (RP).

For LEED 2009 for **Commercial Interiors (CI)**, there are <u>21</u> points for Sustainable Sites (SS), <u>11</u> points for Water Efficiency (WE), <u>37</u> points for Energy and Atmosphere (EA), <u>14</u> points for Materials and Resource (MR), <u>17</u> points for Indoor Environmental Quality (IEQ), <u>6</u> bonus points for Innovations in Design (ID), and <u>4</u> bonus points for Regional Priority (RP).

For LEED 2009 for **Existing Building Operation and Maintenance (EB: O&M)**, there are <u>26</u> points for Sustainable Sites (SS), <u>14</u> points for Water Efficiency (WE), <u>35</u> points for Energy and Atmosphere (EA), <u>10</u> points for Materials and Resource (MR), <u>15</u> points for Indoor Environmental Quality (IEQ), <u>6</u> bonus points for Innovations in Design (ID), and <u>4</u> bonus points for Regional Priority (RP).

LEED for Homes will <u>not</u> be updated until 2011, so its points for various certification levels <u>are still</u> **different** than LEED 2009 or LEED v3.0:

Certified	<u>45</u>–59	points
Silver	<u>60</u>–74	points
Gold	<u>75</u>–89	points
Platinum	<u>90</u>–136	points
<u>Total</u> possible points	**136**	points

For **LEED for Homes,** there are <u>3</u> points for **Innovations and Design Process (ID)**, <u>10</u> points for **Locations and Linkage (LL)**, <u>22</u> points for Sustainable Sites (SS), <u>15</u> points for Water Efficiency (WE), <u>38</u> points for Energy and Atmosphere (EA), <u>16</u> points for Materials and Resource (MR), <u>21</u> points for Indoor Environmental Quality (IEQ), and <u>3</u> points for **Awareness and Education (AE).**

The pilot version of the **LEED for Retail** rating system was released in July 2008. It has not been updated to LEED v3.0, so its points for various certification levels <u>are still</u> **different** than LEED 2009 or LEED v3.0:

Certified	<u>26</u>–32	points
Silver	<u>33</u>–38	points
Gold	<u>75</u>–51	points
Platinum	<u>52</u>–70	points
<u>Total</u> possible points	**70**	points

For the pilot version of **LEED for Retail,** there are <u>16</u> points for Sustainable Sites (SS), <u>5</u> points for Water Efficiency (WE), <u>17</u> points for Energy and Atmosphere (EA), <u>13</u> points for Materials and Resource (MR), <u>14</u> points for Indoor Environmental Quality (IEQ), and <u>5</u> bonus points for Innovations in Design (ID).

The current version of **LEED-NC Application Guide for Multiple Buildings and On-Campus Building Projects (AGMBC)** was released in <u>October 2005</u> and is for use with the LEED-NC Green Building Rating System v2.1 and v2.2. It has <u>not</u> been updated to LEED v3.0 yet, so its points for various certification levels <u>are still</u> **different** than LEED 2009 or LEED v3.0:

Certified	<u>26</u>–32	points
Silver	<u>33</u>–38	points

Gold   <u>75</u>–51 points
Platinum   <u>52</u>-69 points
<u>Total</u> possible points **69** points

For the current version of the **LEED-NC Application Guide for Multiple Buildings and On-Campus Building Projects (AGMBC),** there are <u>14</u> points for Sustainable Sites (SS), <u>5</u> points for Water Efficiency (WE), <u>17</u> points for Energy and Atmosphere (EA), <u>13</u> points for Materials and Resource (MR), <u>15</u> points for Indoor Environmental Quality (IEQ), and <u>5</u> bonus points for Innovations in Design (ID).

4. **How many different levels of building certification does USGBC have?**
   Answer: Four: <u>Certified, Silver, Gold, and Platinum</u>. They are based on the points that a building earns under the LEED green building rating system. For example, based on the checklists provided by the USGBC (See appendixes for links to download these checklists):

   No matter which **LEED v3.0** rating system you choose, each LEED v3.0 rating system has **100** <u>base</u> points; a maximum of **6** possible <u>extra</u> points for Innovation in Design, and a maximum of **4** possible <u>extra</u> points for Regional Priority (RP) for <u>each</u> project. There are 6 possible RP points, but you can only pick and choose a maximum of 4 points for each project.

Certified	**40**–49 points
Silver	**50**–59 points
Gold	**60**–79 points
Platinum	**80** points and above

   See link below to download an Excel file showing the **Regional Priority Credits (RPC)** for all 50 States. With this spreadsheet, you can locate the RPC for your area by <u>zip code</u>:
   http://www.usgbc.org/DisplayPage.aspx?CMSPageID=1984

   A LEED Project team does not have to do anything special, since LEED 2009 Online will automatically decide which RPC your project will get once you enter the project zip code and other information. If you have more than 4 RPCs, then you need to decide which 4 RPCs you want to use for your project.

   The USGBC is working on developing similar RPC incentives for international projects.

5. **What is the process for LEED certification? What are the basic steps for LEED certification?**
   Answer: There are differences between the certification process for LEED for homes and the process for Commercial LEED projects.

   LEED certification **process** for **commercial** projects:
   1) At the **Beginning Stage**:
      a) Value <u>assessment</u>
      b) Use checklists to decide the <u>preliminary LEED score</u>
      c) Select project <u>goals</u>
      d) The documentation/assessment of the recommendations for <u>condition treatment</u>
      e) **Registration**, i.e. register your project online at www.gbci.org (The most <u>important</u> thing in the first step). The project administrator has access to the CIR database and LEED online after registration.

    f) Identification of project <u>partners</u>

    g) Set goals for <u>green building practice</u>

    h) Submit <u>incentives</u> applications

    i) Start the process for <u>documentation</u>

As part of the registration, the LEED project team will have to agree to report <u>post-occupancy water and energy use</u>. This is to assist the USGBC to have a better understanding of the relationship between building performance and LEED credits.

There are some ways to achieve the reporting requirements, including signing up for LEED O&M, OR signing a waiver to allow the USGBC to get the data directly from the utility company.

2) **Design** Stage:

    a) <u>Assemble</u> documentation for the design phase

    b) <u>Design Submittal</u>

    c) USGBC <u>design</u> phase review

    d) No credit will be awarded at the design submittal. The USGBC will notify you if the credit is <u>denied</u> or <u>anticipated</u>. You can omit this stage and submit everything during the Construction Submittal phase.

    e) You need to finish design submittal before beginning work on the construction documents.

3) **Construction** Stage:

    a) Assemble documentation for the Construction phase

    b) Construction Submittal occurs at completion of the project

    c) USGBC <u>construction</u> phase review

    d) Credit is <u>awarded</u> or <u>denied</u>.

    e) Make sure you submit any changes to the design submittal for USGBC's review and approval.

4) **Certification** of your project:

    a) Award Certification

    b) You will receive <u>an award letter, plaque and certificate</u> for your building.

    c) You have <u>30 days</u> to appeal any comments from the USGBC. You need to pay <u>$500</u> for each appeal.

5) **Misc:**

    a) The fees are <u>lower</u> for USGBC members, and the fees can be paid at different stages (Design Submittal and Construction Submittal, etc). You can also pay the fees online.

    b) You can check your project status online.

    c) USGBC will start to review your project <u>10 working days</u> after receiving your submittal.

LEED certification process for **homes**:

1) At the **Beginning** Stage:

    a) Contact the <u>LEED for Home provider</u>

    Note: A LEED for Homes Provider is the referee when it comes to who is able to be a **Green Rater** for a **LEED** for Homes project. The Provider is responsible for hiring, training, and overseeing the Green Raters. The USGBC requires that each Provider have a quality assurance protocol for its Green Raters. Homebuilders who intend to achieve LEED for Homes certification **must contact a LEED for Homes Provider** organization. See link below for a complete list of LEED for Homes Provider organizations in the US: http://www.usgbc.org/DisplayPage.aspx?CMSPageID=1554#CA.

A Green Rater is a professional who has shown qualifications for an energy rating certification, and demonstrates abilities to deal with a home's energy systems and performance. The certifications can be obtained from certification bodies like ENERGY STAR, RESNET, BPI, etc.

  b) Select project goals
  c) **Registration**, i.e. register your project online (The most important thing in the first step). The project administrator has access to the CIR database and LEED online after registration.
  d) Identification of project partners
  e) Submit incentives applications

2) **Design** Stage:
  a) Execute preliminary rating
  b) Organize and submit IDs and CIRs
  c) Organize construction documents (CD)
  d) Set up a durability plan
  e) Set up contracts including the scope of work for various trades

3) **Construction** Stage:
  a) **Green rater** carries out thermal bypass inspection
  b) Green rater carries out optional durability inspection
  c) Various trades sign off on finished tasks
  d) Fill out accountability form

4) **Verification** Stage:
  a) Green rater finishes verification
  b) Provider review of green rater's work
  c) Provider submits documentation to the USGBC

5) **Certification** of you project:
  a) USGBC review
  b) "Denied" or "Achieved"
  c) You have 30 days to appeal any comments from the USGBC. You need to pay $500 for each appeal.

  If **achieved**:
  a) Award Certification
  b) You will receive an award letter, plaque and certificate for your building.

6) **Misc:**
  a) The fees are lower for USGBC members, and the fees can be paid at different stages (Design Submittal and Construction Submittal, etc). You can also pay the fees online.
  b) You can check your project status online.
  c) USGBC will start to review your project 10 working days after receiving your submittal.

6. What does the registration form include?

Answer: Your account login information, project contact, project <u>details</u> (type, title, address, owner, gross square footage, budget, site condition, current project phase, scope, occupancy, etc.) and if the project is confidential or not.

7. What is precertification?

Answer: Precertification is <u>for the Core and Shell program only</u>. It is a formal recognition of a project where the owner intends to seek Core and Shell certification, and gives the owner/developer <u>a marketing tool to attract financers and tenants</u>. A project needs to meet a scorecard requirement for precertification. Precertification is granted after an early design review by the USGBC, and the review usually takes less than <u>one</u> month. Precertification costs $2,500 for USGBC members and $3,500 for non-members. It does <u>NOT</u> cover the fee for certification, and it does <u>NOT</u> guarantee the final certification of the project.

8. What is CIR?

Answer: CIR stands for <u>Credit Interpretation Request</u> or <u>Credit Interpretation Rulings</u>, depending upon its context.

9. When do you submit a CIR?

Answer: When there are <u>conflicts</u> between two credit categories or the USGBC Reference Guide does <u>not</u> give you enough information. The building registration fee used to cover two free CIRs, but after 11/15/05, it costs <u>$200</u> to submit each CIR.

10. What are the steps for submitting a CIR?

Answer:

1) Check your projects against each LEED credit or prerequisite.
2) Check the USGBC Reference Guide.
3) Review the GBCI and USGBC websites for <u>previous</u> CIRs.
4) If you cannot find a similar CIR, then submit a new CIR to GBCI online.
5) Do <u>not</u> mention the <u>contact information, name of the credit, or confidential information</u>.
6) Submit only the essential and required information, and do <u>not</u> submit it in a letter format.
7) Submit <u>one CIR for each prerequisite or credit</u>.
8) Do not include any attachments.
9) Include details and background information, <u>600 words</u> max (4,000 characters including spaces).

11. Will submitting a CIR guarantee a credit?

Answer: No, a CIR only provides feedback, and it will <u>not</u> guarantee a credit. Also, <u>no</u> credit will be awarded in the CIR process.

12. What tasks are handled by GBCI and what tasks are handled by USGBC?

Answer: In 2008, the Green Building Certification Institute (GBCI) spun off from the USGBC.

The USGBC still handles the online tool, LEED rating system development, and related educational offerings, etc.

The GBCI took over some of the responsibilities from the USGBC and handles building LEED certification and LEED professional credentialing.

For building LEED certification, the GBCI oversees 10 organizations including Lloyd's Register Quality Assurance and Underwriters Laboratories, which manage the project review process.

This separation of the tasks for the USGBC and the GBCI will meet the protocols of the American National Standard Institute (ANSI) and International Organization for Standardization (ISO), and will make the building certification a true third-party process.

13. What are MPRs?
Answer: MPRs are LEED's **Minimum Program Requirements**. A project must meet MPRs to qualify for LEED certification.

MPRs serve three goals:
1) Clear guidance for customers
2) Maintain LEED program integrity
3) Make the LEED certification process easier

MPRs include some very basic requirements, like:
1) The building must be 1,000 s.f. minimum for LEED-NC, LEED Schools, LEED-CS and LEED-EB: O&M, and 250 s.f. minimum for LEED-CI.
2) A LEED project must be a permanent and complete space or building.
3) A LEED project must comply with environmental laws
4) A LEED project must have a reasonable site boundary
5) The building to site ratio must be 2% or higher.
6) A LEED project must commit to sharing water and whole-building energy usage data
7) A LEED project must meet a minimum occupancy rate

See link below for detailed information:
http://www.usgbc.org/projecttools

14. Which types of projects should use LEED-NC?
Answer: Use LEED-NC for new construction or major renovation for commercial occupancies, like a residential building of 4 or more habitable floors, hotels, offices, or institutional buildings (churches, museums, libraries).

A **major renovation** means major building envelope changes, significant HVAC renovation and interior rehabilitation.

If a building project does not involve major renovation, then it is better to seek LEED-EB: O&M certification.

15. Which types of projects should use LEED-CS?
Answer: Use LEED-CS for projects where the developer controls the core and shell (building envelope, HVAC and fire protection systems), but has no control over tenant improvements. These projects may include retail centers, lab facilities, warehouses, medical office buildings, or commercial office buildings. If the developer occupies **50% or less** of the building's **leasable area**, then he can also seek LEED-CS certification.

16. Which types of projects should use LEED for Schools?

Answer: Use LEED for Schools on new construction or <u>major renovation</u> of school buildings. Academic buildings in K-12 schools <u>must</u> use LEED for Schools.

Prekindergarten and post-secondary academic buildings can use either LEED-NC or LEED for Schools.

Non-academic campus buildings like dormitories, maintenance facilities or administrative buildings can use either LEED-NC or LEED for Schools.

If a campus project does not involve new construction or <u>major renovation,</u> then LEED-EB: O&M is often most applicable.

Chapter 4
LEED Green Associate Exam
Technical Review

In this chapter, we introduce the LEED certification process, key components of the LEED rating system, as wells as the purpose, core concepts, strategies, incentives, recognitions and regulations for each LEED credit category.

1. **What do green buildings address?**
   Green buildings mitigate degradation of ecosystems/habitat and resource depletion, reduce costs of operating and owning living and work spaces, improve occupant productivity and comfort, indoor environmental quality, and reduce water consumption.
   Mnemonics: DR. COIN C. See underlined letters in sentences above.

2. **Key stake holders and an integrative approach**
   Green buildings employ an **integrative approach** and encourage the participation of **key stake holders**, including the:
   1) **Client:** Facilities Management Staff, Facilities O&M Staff, Community Members, Owner, and Planning Staff. **Mnemonics**: MOM COP

   2) **Design team:** Civil Engineer, Landscape Architect, Architect, Mechanical and Plumbing Engineer, Electrical Engineer, Structural Engineer, Commissioning Authority, and Energy and Daylighting Modeler.
   Mnemonics: CLAMPES CAM

   3) **Builder:** General Contractors, EMP Subcontractors, Cost Estimator, Construction Manager, and Product Manufacturer.
   Mnemonics: G EMP EMP

   There are many **benefits** of an **integrative approach**, including better indoor air quality, improved occupant performance, reduced operating and maintenance costs, durable facilities, reduced environmental impact, potentially no increase in construction cost, optimized return on investment, and opportunity of learning.
   Mnemonics: bird in pool

3. **The Mission of USGBC**
   The **mission** of the USGBC is:
   1) To improve the quality of life
   2) To change the way we design, build and operate communities and buildings
   3) To create a healthy, prosperous, socially and environmentally responsible environment

4. **The structure of LEED rating system**

   In addition to the main credit categories of SS, WE, EA, MR, IEQ, ID and RP, the following factors are very important to the LEED rating systems: linkages and locations, awareness and education, green buildings and infrastructure, neighborhood design and pattern.

   The structure of the LEED rating system includes three tiers:

   1) The main credit categories of SS, WE, EA, MR, IEQ, ID, and RP
   2) Each main credit category include prerequisites and credits
   3) Each prerequisite or credit includes intents and requirements (or paths to achieve LEED points)

5. **LEED certification tools**

   LEED certification tools include:
   1) USGBC Reference Guides
   2) LEED Rating systems
   3) LEED Online
   4) LEED Scorecard
   5) LEED letter template
   6) CIRs (Credit Interpretation Rulings)
   7) LEED Case Studies
   8) USGBC and GBCI websites:
      www.USGBC.org
      www.GBCI.org

   The GBCI requires that the project team submit an overall narrative and completion of the LEED Online documentation for the LEED certification application. The general documentation includes project timeline and scope, site conditions, project team identification, usage data and occupant, etc.

   The project's overall narrative includes the team, building, site, and applicant's organization.

   LEED Online includes the project's detailed information and template/complete documentation requirements for completing prerequisites and credits: my Action Items, potential LEED ratings, attempted credit summary (not awarded, anticipated, denied, and total attempted), appealed credit summary, and credit scorecard, etc. It also includes embed tables and calculators to assure accuracy and completeness of the submittal package.

   LEED Online also includes definition for **declarant**, Licensed Professional Exemption Form and related information. A **declarant** is the team member(s) who signs off on the documents and indicates who is responsible for each credit or prerequisite.

   A licensed professional exemption form is the form for a team member who is a licensed/ registered landscape architect, architect or engineer to use as a tool to request waiver for eligible submittal requirements. Licensed professional exemptions are shown in the related LEED credit section of LEED Online.

   With LEED Online, project teams can upload support files, submit applications for reviews and CIRs, receive feedback from the reviewer, contact customer service, generate specific reports for project, and obtain project LEED certification, as well as gain access to additional LEED resources like tutorials, FAQs, sample documentation, offline calculators.

If you have <u>multiple projects</u>, you can access all of them via LEED Online.

The GBCI also issues LEED <u>certificates</u> for successful projects via LEED Online.

The credit scorecard includes the project name, address and a list of all points in various credit categories. With LEED Online, you can chose to collapse all credit categories or view a printer friendly scorecard.

The LEED letter template includes your name, your company name, specific and credit-related project information like project site area (s.f.), gross building area (s.f.), and the credit path you choose for each credit.

As a LEED professional, you also need to know the **Synergies** between main credit categories of <u>SS, WE, EA, MR, IEQ, ID, and RP</u>. Almost every LEED exam will have a large percentage of test questions related to **Synergies.**

6. Specific Technical Information

A. Sustainable Sites (SS)

Overall purpose:
1) <u>C</u>**onstruction activity pollution prevention:** prevent soil erosion by wind or storm water runoff during construction; protect and stockpile topsoil for reuse; avoid waterway sedimentation; prevent airborne dust and particle matter from polluting the air.
2) <u>S</u>**ite selection: develop proper sites:** locate a building wisely on a site to alleviate environmental impact.
3) <u>D</u>**evelopment density and community connectivity: reuse existing sites and/or buildings:** encourage development in urban areas with existing infrastructure.
4) <u>B</u>**rownfield redevelopment: restore and/or protect sites:** rehabilitate damaged sites with environmental contamination; restore dam-aged habitats to promote biodiversity.
5) <u>A</u>**lternative transportation: reduce automobile use:** reduce land development impact and pollution.
6) <u>S</u>**ite development: protect or restore habitat and maximize open space; protect prime farmland and undeveloped land/natural areas** (habitat and greenfields) and natural resources; promote biodiversity via a high ratio of open space to development footprint.
7) <u>S</u>**torm water design: quality control and quantity control:** manage storm water runoff; increase on-site infiltration; reduce impervious cover; eliminate contaminants; eliminate storm water run-off pollution; limit disruption of natural water hydrology/natural water flows; capture and process storm water runoff with the help of a storm water management plan and acceptable **BMPs (best management practices).**
8) <u>H</u>**eat island effect: non-roof and roof:** reduce heat islands' impact on wildlife habitat, humans and microclimate. Heat island refers to the extra thermal gradient in developed areas when compared with undeveloped areas.
9) <u>L</u>**ight pollution reduction:** reduce development impact on nocturnal environments; reduce glare and improve nighttime visibility, and reduce sky-glow and light trespass from site and building to increase night sky access.

Mnemonics: Carole Smith Does Business As "So Sweet, Hot Love." (See underlined letters above also).

Core concepts:
1) **Location/Transportation**
   a) Reduce demand for transportation
   b) Improve the efficiency of transportation
2) **Selection of site**
   a) Develop in existing high density areas
   b) Brownfield redevelopment
   c) Connectivity and site selection
3) **Management and design of the site**
   a) Stewardship of the site
   b) Site development
   c) Light pollution reduction
   d) Pest management integration
4) **Management of storm water**
   a) Storm water quantity reduction and water quality protection
   b) Impervious surface impact

Recognition, regulation and incentives:
1) **Financial incentives**
   Local, State and Federal government programs
   - Encourage the reuse of "brownfield" sites or infill
   - Promote water quality protection with the help of combining low impact development and smart growth

Overall strategies and technologies:
Note: Not **all** strategies and technologies have to be used simultaneously for your project.
1) Development density (use existing infrastructure) OR
2) Community connectivity:
   - Consider diverse land uses and provide pedestrian access
   - Provide access to recreational and public spaces
   - Plan walkable, safe connections between buildings and surrounding developments and open spaces
3) Develop in previously developed locations
4) Use telecommuting and flexible work schedules to reduce transportation demand
5) Support efficient modes of transportation
6) Provide access to transit
7) Provide bicycle accessibility
8) Use efficient vehicles
9) Support car/van pools
10) Site development: protect or restore habitat
11) Restore and/or protect open spaces:
    Plan for protected areas on-site and plan for easements and protected areas off site
12) Manage and intercept storm water
13) Apply cool roof technologies
14) Reduce duration of lighting use and lighting density, also use light fixtures that comply with dark

sky requirements

Specific Technical Information:

Note: The USGBC workshops may not give you enough information to pass the LEED Green Associate Exam, so I have included <u>part</u> of the specific LEED 2009 BD&C technical information to fill in these blanks. This is only a <u>partial</u> version (about 70%) of the information based on LEED 2009 BD&C. Most of the specific information that follows is common or useful to <u>ALL</u> LEED systems, and I think the information will help you.

For the LEED Green Associate Exam, based on my research and the USGBC workshops that I have attended, it is unlikely that the GBCI will require you to differentiate each <u>detailed</u> credit, i.e., whether the information is in SSc1 or SSc5.1 and how many exact points SS5.1 is worth, but the GBCI will want you to know that the information is contained within the <u>main</u> category of SS, and you DO need to know <u>some</u> technical information for the exam. You also definitely need to know how many points your project will need to become LEED-Certified, LEED-Silver, LEED-Gold and LEED-Platinum, and the total possible number of ID points, RP points, etc.

There are a few detailed credits for school or CS <u>only</u>. I do <u>not</u> include these because the LEED Green Association Guide is unlikely to test a credit with specialized information for schools or CS only.

Some credits are for **NC, CS,** and **Schools** and they do have some <u>generic or common</u> LEED <u>technical</u> information with some <u>minor</u> variations between **NC, CS,** and **Schools**. The GBCI may test these, so I have included them in the book.

I still present the information in the detailed credit format (SSp1, SS5.1, etc.), because it helps to organize the information in a useful way and makes it easier to understand. If I presented all the information for each main category altogether without breaking it down, it would be 60 or 80 items for each main credit category. It would be confusing and difficult for you to understand and digest.
It is also better to know more than what the exam will test rather than to know less and end up failing.

Here is the specific technical information for SS:

SSp1 (SS Prerequisite 1): Construction Activity Pollution Prevention (<u>Required</u> for NC and CS and Schools)

Purpose:
To control <u>airborne</u> dust, <u>soil</u> erosion and <u>waterway</u> sedimentation from construction sites.

Credit Path:
<u>Implement an **Erosion and Sedimentation Control (ESC)** Plan.</u> The ESC Plan shall comply with the <u>2003 EPA **Construction General Permit (CGP)** or local codes</u>, whichever is more restrictive.

The objectives are:
1) Prevent soil erosion by wind or storm water runoff during construction.
2) Protect and stockpile topsoil for reuse.
3) Avoid sedimentation.
4) Preventing dust and particulate matter from polluting the air.

The CGP describes the requirements needed to conform to Phase I and Phase II of the **National Pollutant Discharge Elimination System (NPDES)** program. The CGP only applies to construction sites larger than one acre, but the LEED requirements apply to all projects for this prerequisite's purpose.

Submittals:
1) A sedimentation and erosion control plan/drawing.
2) A narrative detailing **BMP** and **ESC** and responsible parties.
3) Confirmation showing compliance with local erosion control standards or **NPDES**.
4) Reports or inspection logs, date-stamped photos, etc.

Synergies:
Site restoration efforts, reducing site disturbances, and preventing sedimentation and erosion will contribute to:
- SSc5.1: Site Development: Protect or Restore Habitat
- SSc5.2: Site Development: Maximize Open Space

Low impact development strategy, limiting the disruption of natural hydrology will contribute to:
- SSc6.1: Storm Water Design: Quantity Control
- SSc6.2: Storm Water Design: Quality Control

Possible Strategies and Technologies:
1) Create an ESC plan early in the project, preferably during the design stage.
2) Use the Stabilization Method: mulching, temporary or permanent seeding.
3) Use Structural Method if erosion has already occurred: silt fencing, earth dikes, sediment basins and sediment traps.

Extra Credit (Exemplary Performance): None

Project Phase: Schematic Design (**Note:** We separate a project into the following phases: Pre-Design, Schematic Design, Design Development, Construction Documents, Construction Administration, and Occupation/Operation.)

LEED Submittal Phase: Construction
(**Reminder:** Some LEED credits or prerequisites may be submitted during the Design Submittal phase while others can ONLY be submitted during the Construction Submittals phase.)

Related Code or Standard:
2003 EPA Construction General Permit (CGP)
National Pollutant Discharge Elimination System (NPDES)

Responsible Party: Civil Engineer and Contractor

SSp2: Environmental Site Assessment (<u>Required</u> for Schools Only)

Note: Detailed discussions have been omitted since this prerequisite is for schools only, and is unlikely to be tested on the LEED Green Associate Exam.

SSc1: Site Selection (<u>1</u> Point for NC and CS and Schools)

Purpose:
1) To avoid inappropriate sites.
2) To wisely locate a building on a site to alleviate environmental impact.

Credit Path:
Avoid sites that meet any one of the criteria below:
   a) Virgin (previously undeveloped) land that is <u>lower than five feet above</u> the elevation of the <u>100-year flood</u> per definition of Federal Emergency Management Agency (<u>FEMA</u>).
   b) Virgin land that is within <u>50 feet</u> of a water body, such as a lake, river, stream or sea (<u>NOT</u> including small man-made ponds), per the terminology of the <u>Clean Water Act</u>.
   c) Within <u>100 feet of wetlands</u> per definition of the US <u>Code of Federal Regulations</u>, and areas of special concern or isolated wetlands per local rules, <u>OR</u> within setback distance from wetlands per local codes, whichever is more restrictive.
   d) <u>Habitat</u> for any species on state <u>OR</u> federal endangered or threatened lists.
   e) <u>Public parkland</u> unless a public land owner has accepted a trade for land of equal or greater value as parkland (This does <u>not</u> apply to Park Authority projects).
   f) <u>Prime farmland</u> per the definition of the <u>US Department of Agriculture (USDA)</u> in the <u>US Code of Federal Regulations (CFR)</u>.

Submittals:
1) Copies of <u>summaries</u> from all ASTM assessment performed
2) If remediation is needed, submit a description of the <u>efforts</u>.
3) Documentation from <u>governing agencies</u> (federal EPA region, state and local) showing completion of remediation to standards for unrestrictive residential use.

Synergies:
Site remediation <u>efforts</u> and Phase I and II environmental <u>assessment</u> will contribute to:
   • SSc3: Brownfield Redevelopment

Possible Strategies and Technologies:
1) Avoid restrictive land types and sensitive site elements listed above.
2) Pick a proper building location.
3) Design the building with the minimum possible footprint to avoid the sensitive elements listed above.

Extra Credit (Exemplary Performance): None

Project Phase: Pre-Design

LEED Submittal Phase: <u>Design</u>

Related Code or Standard:
1) <u>FEMA</u>
2) <u>USDA</u>
3) <u>CFR</u>
4) <u>Clean Water Act</u>
5) U.S. Fish and Wildlife Service, <u>List of Threatened and Endangered Species</u>
6) National Marine Fisheries Service, <u>List of Endangered Marine Species</u>

Responsible Parties: <u>Civil Engineer</u> and Owner

SSc2: Development Density and Community Connectivity (<u>5</u> Points for NC and CS, <u>4</u> Points for Schools)

Purpose:
1) To protect natural resources, habitat and greenfields.
2) Encourage development in urban areas with existing infrastructure.

Credit Paths:
1st Path: Development Density
Renovate or construct the building on a developed site AND in an area with at least <u>60,000 s.f. per acre</u> net density. You must include the proposed project area and base the density calculation on a two-story downtown development.

OR
2nd Path: Community Connectivity
Renovate or construct the building on a developed site AND within ½ mile of at least <u>10</u> Basic Service AND within ½ mile of a neighborhood or residential area with at least <u>10 units per acre</u> net density AND with pedestrian access between the service and the building. Note: For Schools, the ½ mile radius can be drawn <u>from any building entrance</u> and counting the services inside the radius. For CS and NC, the ½ mile radius can be drawn <u>from a main building entrance</u> and can include the services within the radius.

The qualified Basic Services include, but are not limited to:
a) Place of Worship;
b) Restaurant;
c) Supermarket;
d) Convenience Grocery;
e) Laundry;
f) Cleaner;
g) Beauty Salon;

    h) Hardware;
    i)   Pharmacy;
    j)   Medical/Dental;
    k) Bank;
    l)   Senior Care Facility;
    m) Community Center;
    n) Fitness Center;
    o) Daycare;
    p) School;
    q) Library;
    r)   Museum;
    s)   Theater;
    t)   Park;
    u) Fire Station;
    v) Post Office.

Up to 2 of the 10 basic services can be anticipated.

You can only count <u>one</u> of each service, except <u>two</u> restaurants may be counted.

Submittals:
For 1st Path
    1) A site <u>vicinity plan</u> highlighting the development <u>density</u>
    2) <u>Records</u> of building and site <u>area</u>

For 2nd Path
    1) A site <u>vicinity plan</u> highlighting locations of the <u>residential</u> areas, the ½ <u>mile radius</u>, and the types and locations of qualifying <u>services</u>.

Synergies:
Channeling development to previously developed urban areas near public transportation may contribute to:
- SSc1: Site Selection
- SSc4.1: Alternative Transportation

Possible Strategies and Technologies:
Select urban sites with pedestrian access to basic services listed above.

Extra Credit (Exemplary Performance):
You can get <u>one</u> innovation point by <u>doubling</u> the requirements, i.e. 120,000 s.f per acre.

Project Phase: Pre-Design

LEED Submittal Phase: <u>Design</u>

Related Code or Standard:
None

Responsible Parties: <u>LEED AP</u> and Owner

SSc3: Brownfield Redevelopment (1 Point for NC and CS and Schools)

Purpose:
1) To save undeveloped land.
2) Rehabilitate damaged sites with environ-mental contamination.

Credit Paths for NC and CS:
1) Choose and remediate a contaminated site as documented by a local voluntary cleanup program or by an ASTM E1903-97 Phase II Environmental Site Assessment;

OR
2) A brownfield as defined by a federal, state or local governing agency.

Credit Paths for Schools:
The credit can only be achieved by SSp2: Environmental Site Assessment and remediating site contamination.

Note: Per USGBC LEED Addenda (http://www.usgbc.org/addenda), a project where asbestos is found and remediated can earn this credit. See link below for information on asbestos:
http://www.epa.gov/asbestos/index.html

Submittals:
1) For NC and CS, a description of site contamination and remediation efforts
2) For schools, executive summaries from ASTM site assessment performed

Synergies:
Projects developed on brownfield sites may contribute to:
- SSc1: Site Selection

Possible Strategies and Technologies:
1) Select brownfield sites.
2) Coordinate remediation activity and site development plans.
3) Identify property cost savings and tax incentives.
4) Pump-and-Treat; **In-situ (on-site)** remediation: injection wells or reactive trenches; **ex-situ (off-site)** remediation: bioreactors, chemical treatment; land farming; bioremediation: use micro-organization and vegetation.

Extra Credit (Exemplary Performance): None

Project Phase: Pre-Design

LEED Submittal Phase: Design

Related Code or Standard:
1) ASTM E1527-05 Phase I Environmental Site Assessment

2) <u>ASTM E1903-97</u> Phase II Environmental Site Assessment, effective 2002
3) Local Codes
4) U.S. <u>EPA</u>, Definition of Brownfields
5) <u>CERLA</u>: Comprehensive Environmental Response, Compensation and Liability Act (Super Fund)

Responsible Party: <u>Civil Engineer</u> and Owner

SSc4.1: Alternative Transportation: Public Transportation Access (<u>6</u> Points for NC and CS, <u>4</u> points for Schools)

Purpose:
To reduce automobile use, reduce land development impact and pollution.

Credit Paths for NC, CS and Schools:
1) The project shall be located within ½ mile of <u>one</u> planned and funded, or <u>one</u> existing, commuter <u>rail</u> or subway station.

OR
2) The project shall be located within ¼ mile of at least one stop for <u>two</u> or more campus or public <u>bus</u> lines usable by the occupants of the building.

Note: All distances above shall be measured from a <u>main</u> building entrance. You can only count a school's bus system as <u>one</u> bus line.

Credit Paths for Schools Only:
3) **Pedestrian Access:** <u>80%</u> of the students in Grade <u>8</u> or below live within a <u>3/4</u>-mile walking distance of the school, and <u>80%</u> of the students in Grade <u>9</u> and above live within a 1 1/2-mile walking distance of the school. Provide pedestrian access from the students' homes to school.

For all three paths:
Provide dedicated bike lanes and walking paths to transit lines; and provide bike lanes extending in two or more directions from the school property line with no barriers (fences, etc.).

Submittals:
1) Identify <u>bus</u> routes and <u>rail stations</u> serving the project
2) Create a site <u>vicinity map (to scale)</u> with the <u>walking paths</u> between bus stops, rail stations, and the project building's main entrance <u>labeled</u>

Synergies:
Projects close to public transportation tend to be in previously developed and densely developed areas. They may qualify for:
    SSc1: Site Selection
    SSc2: Development Density and Community Connectivity
Possible Strategies and Technologies:

1) Locate the building close to mass transit.
2) Survey future building occupants to determine their transportation needs.

Extra Credit (Exemplary Performance):
You can get <u>one</u> innovation point by <u>doubling</u> the requirements, i.e., <u>two</u> train stations or <u>four</u> bus lines.

Project Phase: Pre-Design

<u>LEED Submittal</u> Phase: <u>Design</u>

Related Code or Standard: None

Responsible Party: <u>LEED AP</u> and Owner

SSc4.2: Alternative Transportation: Bicycle Storage and Changing Rooms (<u>1</u> Point for NC and Schools, <u>2</u> points for CS)

Purpose:
1) To reduce automobile use.
2) To reduce land development impact and pollution.

Credit Paths for NC:
1) For <u>institutional or commercial</u> buildings: Within <u>200 yards</u> of the building entrance, provide secure bicycle storage/rack for at least <u>5%</u> of the <u>peak</u> period building occupants, AND provide changing rooms and showers for <u>0.5%</u> of **Full-Time Equivalent (FTE)** users within <u>200 yards</u> of the entrance for the building, or in the building. Note: <u>FTE Occupants=Occupants Hours/8</u>

    OR
2) For <u>residential</u> buildings, provide secure and covered bicycle storage facilities for at least <u>15%</u> of the building users instead of shower/changing rooms.

Credit Paths for CS:
1) For <u>institutional or commercial</u> buildings with 300,000 s.f. or less, within <u>200 yards</u> of the building entrance, provide secure bicycle storage/rack for at least <u>3%</u> of the building's occupants (Based on the <u>average</u> for the year), AND provide changing rooms and showers for <u>0.5%</u> of **Full-Time Equivalent (FTE)** users within <u>200 yards</u> of the entrance for the building, or in the building. Note: <u>FTE Occupants=Occupants Hours/8</u>

2) For <u>institutional or commercial</u> buildings larger than 300,000 s.f., within <u>200 yards</u> of the building entrance, provide secure bicycle storage/rack for at least <u>3%</u> of the building's occupants (based on <u>average</u> for the year) for up to 300,000 s.f., and an additional 0.5% for the building's occupants for space over 300,000 s.f., AND provide changing rooms and showers for <u>0.5%</u> of the **Full-Time Equivalent (FTE)** users within <u>200 yards</u> of the entrance for the building, or in the building.

You must apply this rule for <u>each</u> building <u>use type</u> on mixed-use buildings larger than 300,000 s.f.

OR

3) For <u>residential</u> buildings, provide secure and covered bicycle storage facilities for at least <u>15%</u> of the building's users instead of shower/changing rooms.

Note: For all three paths above, see Appendixes for "Default Occupancy Factors" Table.

Credit Paths for Schools:
1) Within <u>200 yards</u> of the entrance for the building, provide secure bicycle storage/rack for at least <u>5%</u> of the <u>peak</u> period building <u>staff and students above Grade 3</u>, AND provide changing rooms and showers for <u>0.5%</u> of the **Full-Time Equivalent (FTE)** <u>staff</u> within <u>200 yards</u> of the entrance for the building, or in the building.

Provide bike lanes extending two or more directions from the school property lines with no barriers (fences, etc.).

Submittals:
1) Calculate the <u>number of occupants</u> for each use type and determine the <u>number of showering facilities and bicycle storage</u> needed
2) Create a plan showing the <u>quantity and locations</u> of showering and bicycle storage facilities, also determine the <u>distance</u> between the project building's entry and the facilities

Synergies:
Paving materials for on-site bicycle lanes may affect the following credits:
- SSc6: Storm Water Design
- SSc7.1: Heat Island Effect: Non-Roof

Possible Strategies and Technologies:
Design the building with
1) Shower/changing rooms and/or
2) Bicycle storage/rack.

Extra Credit (Exemplary Performance):
You may get one extra point for SSc4 by implementing a comprehensive program that results in <u>quantifiable</u> automobile use reduction.

Project Phase: Schematic Design

<u>LEED Submittal</u> Phase: <u>Design</u>

Related Code or Standard:
None

Responsible Party: <u>Architect and LEED AP</u>

SSc4.3: Alternative Transportation: Low Emitting and Fuel Efficient Vehicles (3 Point for NC and CS, 2 points for Schools)

Purpose:
1) To reduce automobile use.
2) To reduce land development impact and pollution.

Credit Paths:
1) For **NC,** provide fuel efficient and low emitting <u>vehicles</u> AND related preferred parking for <u>3%</u> of the Full-Time Equivalent (FTE) users.

OR
2) For **NC and CS**, 5% of the total parking is to be designated as preferred parking for fuel efficient and low emitting vehicles. Providing a minimum 20% parking rate discount for fuel efficient and low emitting vehicles to ALL (not just the 5%) customers is an acceptable alternative. The discount rate shall be posted at the parking facility entrance for at least <u>2</u> years.

For **CS**, GBCI will consider alternatives on a case-by-case basis for projects having difficulties complying with preferred parking, defined as closest to main entrance.

For **schools**, 5% of the total parking is to be designated as preferred parking for fuel efficient and low emitting vehicles, AND a minimum of <u>1</u> designated carpool drop-off area for fuel efficient and low emitting vehicles.

OR
3) For **NC and CS**, provide an alternative-fuel <u>refueling station</u> for <u>3%</u> of the total vehicle parking capacity on the site (gas or liquid fueling facilities shall be located outdoors or have separate ventilation).

For **schools**, design and execute a <u>plan</u> for buses and maintenance vehicles serving the school to be fuel efficient and low emitting vehicles, OR to use <u>20%</u> biodiesel, propane, or natural gas.

4) For **NC,** provide building occupants <u>access to</u> a vehicle sharing <u>program</u> utilizing fuel efficient and low emitting vehicles. The following criteria must be met:

- Submit a <u>narrative</u> describing the vehicles-sharing program and its administration
- Provide a <u>2-year-minimum</u> vehicle-sharing <u>contract</u>
- Provide <u>documentation</u> supporting the estimated number of occupants served per vehicle
- Parking for fuel efficient and low emitting vehicles shall be in the <u>closest spaces</u> in the <u>parking lot closest to the main entrance</u>. Submit an area map or site plan with the path from the project site to the parking facility clearly highlighted.
- Provide one fuel efficient and low emitting vehicle for every 3% of the FTE occupants. Assuming eight people can share one vehicle, this means providing one vehicle for every 267 FTE occupants. Provide one fuel efficient and low emitting vehicle if a project has fewer than 267 FTE occupants.

Only vehicles classified as **Zero Emission Vehicles (ZEV)** by California Air Resources Board or vehicles with a green score of at least <u>40</u> on the **American Council for an Energy Efficient Economy (ACEEE)** annual vehicle rating guide are qualified as fuel efficient and low emitting vehicles for this credit.

"Preferred parking" means the parking stalls closest to main building entrance (<u>not including</u> handicap parking) or parking passes at a discounted price.

Submittals:

For 1st Path (NC)
1) Calculate number of FTE building occupants and determine the <u>number of qualifying vehicles</u> which must be provided
2) <u>Documentation</u> about vehicles purchased including make, model and fuel type
3) Create a <u>site plan</u> showing the location of the <u>preferred</u> parking spaces

For 2nd Path
1) For designated spaces, <u>record</u> the number of on-site parking spaces, <u>identify</u> the <u>preferred</u> spaces and <u>inform</u> building occupants
2) For discounted spaces, gather information about the <u>discount program</u> and the <u>means to inform</u> building occupants

For 3rd Path (NC& CS)
1) <u>Documentation about</u> the number, type, manufacturer, model number, and fueling capacity of the <u>fueling stations</u> provided

For 4th Path
1) <u>Information</u> about the fuel efficient and low emitting <u>shared vehicles</u> including quantity, make, model and fuel type
2) A copy of the contracts for the vehicle sharing agreement
3) Information on the <u>vehicle sharing program</u> like estimated <u>number</u> of customers per vehicle and the <u>administration</u> of the program.
4) An area map or <u>site plan</u> highlighting the pedestrian <u>walkway</u> from the project site to the parking area.

For Schools
1) Calculations showing compliance with the 20% alternative fuel requirement, or the 20% low emitting vehicles requirement
2) <u>Information on</u> the fuel efficient and low emitting or alternative-fuel <u>vehicle program</u>

Synergies:
If you provide preferred parking without increasing overall parking capacity, then your project may qualify for:
- SSc4.4: Alternative Transportation: Parking Capacity

Possible Strategies and Technologies:
1) Install an alternative-fuel refueling station and
2) Try to share the cost with neighbors

Extra Credit (Exemplary Performance):
You may get one extra point for SSc4 by implementing a comprehensive program that results in <u>quantifiable</u> automobile use reduction.

Project Phase: Schematic Design

<u>LEED Submittal</u> Phase: <u>Design</u>

Related Code or Standard: None

Responsible Party: <u>Architect, MEP Engineer, LEED AP</u>, and Owner

SSc4.4: Alternative Transportation: Parking Capacity (<u>2</u> points for NC and CS and Schools)

Purpose:
1) To reduce automobile use.
2) To reduce land development impact and pollution.

Credit Paths for NC and CS:
1) For **non-residential** projects, design the parking capacity to <u>meet, but not to exceed, minimum</u> local zoning requirements.

   <u>Additional</u> requirement for **NC**: provide <u>5%</u> preferred parking (based on the total parking capacity of the site) for vanpools or carpools.

OR

2) For **non-residential** projects, if there are no local code requirements, then you need to install <u>25%</u> less parking than what is required by the Institute of Transportation Engineer's (**ITE**) Parking Generation Studies, 3rd Edition. Refer to the link at:

   http://www.ite.org

OR

3) For **non-residential** projects, if your project has parking for less than 5% (for **NC**) or 3% (for **CS**) of the FTE building occupants, you can designate and mark 5% (for **NC**) or 3% (for **CS**) of the total parking as preferred parking (based on the total parking capacity of the site) for vanpool or carpools.

   Providing a minimum 20% parking rate discount for fuel vanpool or carpools vehicles to ALL (not just the 5% or 3%) customers is an acceptable alternative. The discount rate shall be posted at the parking facility entrance for at least <u>2</u> years.

OR

4) For **residential** projects, design the parking capacity to meet, but <u>not to exceed, minimum local</u>

zoning requirements AND provide facilities to support programs encouraging vehicle-sharing, like car-sharing services, shuttle services to mass transit, a ride board, designated parking for vanpools, or carpool drop-off areas, etc.

5) For **mixed-use** projects with 10% or less commercial area, you must follow the requirements for **residential** projects above; For mixed-use projects with more than 10% commercial area, you must follow the requirements for **residential** projects above for the residential portion, and you must follow the requirements for **non-residential** projects above for the commercial portion.

OR

6) For **all** uses, NO new parking.
Additional requirements for **CS:** for **all** paths above, follow the Appendixes for "Default Occupancy Factors" Table.

Credit Paths for Schools:
a) Design parking capacity to meet, but not exceed, minimum local zoning requirements, AND designate 5% of the total parking capacity of the site as preferred parking for vanpool or carpools.
b) If there are no local code requirements, then you need to install 25% less parking than what is required by the Institute of Transportation Engineer's (**ITE**) Parking Generation Studies, 3rd Edition. See link at:
http://www.ite.org
c) NO new parking

Submittals:
1) Information on type and amount of parking, and programming and/or infrastructure supporting vanpool and carpool like number of FTEs and preferred parking spaces, zoning requirements, or brochures for occupants explaining the vanpooling and carpooling support structure

Synergies:
Meeting but NOT exceeding the zoning requirements, and minimizing parking, especially surface parking can contribute to the following credits:
- SSc5.1: Site Development: Protect or Restore Habitat
- SSc5.2: Site Development: Maximize Open Space
- SSc6: Storm Water Design
- SSc7.1: Heat Island Effect: Non-Roof

Possible Strategies and Technologies:
1) Encourage shared vehicle use.
2) Share parking with neighboring buildings.
3) Minimize parking lot size.

Extra Credit (Exemplary Performance):
You may get one extra point for SSc4 by implementing a comprehensive program that results in quantifiable automobile use reduction.

Project Phase: Schematic Design

LEED Submittal Phase: Design

Related Code or Standard:
1) Institute of Transportation Engineer's Parking Generation Studies, 3rd Edition.

Responsible Party: Civil Engineer and Owner

SSc5.1: Site Development: Protect or Restore Habitat (1 Point for NC and CS and Schools)

Purpose:
1) To protect existing habitats and
2) To restore damaged habitats to promote biodiversity.

Credit Paths for NC and CS and Schools:
1) For greenfield sites, limit construction disturbance to
   - 10 feet beyond patios, walkways, surface parking and utilities that are 12 inches in diameter or less
   - 15 feet beyond trenches for utility branches and main roadway curbs
   - 25 feet beyond permeable constructed surfaces like playing fields, storm water detention facilities and pervious paving, which requires staging areas to limit compaction in constructed areas

   OR
   - 40 feet beyond the building boundary

   OR

2) On previously graded or developed land, protect or restore at least ½ of the site area (NOT including the building footprint), OR 20% of the total site area (including the building footprint) if it is larger, with adapted or native plants. **Adapted/native plants** are local plants or cultivars of native plants that are not noxious weeds or invasive species.

   If you are already earning credit for SSc2 and you are using plants on roof surfaces, you can still apply the vegetated roof surface to this calculation if the plants are native or adapted plants, encourage biodiversity, and provide habitat.

   OR

3) If a project has limited landscape opportunities, the project team can permanently donate offsite land, equal to 60% of the previously developed area (including the building footprint), to a land trust located in the same EPA Level III Ecoregion identified for the project site. The land trust must meet the requirements of the Land Trust Alliance 'Land Trust Standards and Practices' 2004 Revisions.

   Greenfield sites are the ones that are not graded or developed previously, and are still in their natural state. Previously developed sites are ones that have been altered by man, were previously graded, or already have parking lots, roadways, or buildings.

Submittals:
1) **For greenfield sites:** Provide a site drawing showing boundaries of disturbance

2) **For previously developed sites:**
   Provide site drawing showing the area to be restored or protected.

Synergies:
Protecting or restoring habitat and use of native vegetation as part of a vegetated roof or on-site landscape features may contribute to the following credits:
- SSc5.2: Site Development: Maximize Open Space
- SSc6.1: Storm Water Design: <u>Quantity</u> Control
- SSc6.2: Storm Water Design: <u>Quality</u> Control
- SSc7.1: Heat Island Effect: Non-Roof
- SSc7.2: Heat Island Effect: Roof
- WEc1: Water Efficiency Landscaping

Possible Strategies and Technologies:
For <u>greenfield</u> sites:
a) Conduct a <u>site elements survey</u> (ecosystems, existing water body, soil conditions, trees and other vegetation, etc.) to find out if sensitive habitats exist on-site. Also, develop a <u>master plan</u> for the project.
b) Locate the building on the site carefully and <u>minimize its footprint</u> to avoid disturbing the existing ecosystem.
c) <u>Share facilities</u> with neighbors.
d) Build <u>multi-story buildings</u> and place parking under the buildings.
e) <u>Restore</u> damaged sites and mark construction boundaries clearly to have a minimum impact on the existing sites.

On <u>previously graded or developed</u> land:
a) Take advantage of <u>local plant societies</u>, consultants, educational facilities, regional and local governing agencies, and seek their assistance in choosing appropriate adapted and native plants.
b) <u>Ban noxious weeds and invasive species</u>.
c) Use <u>low maintenance, drought-resistant</u> plants.
d) <u>Avoid the use of herbicides</u>, pesticides and fertilizers and monoculture plantings, and promote biodiversity and habitat value.

Extra Credit (Exemplary Performance):
You can get <u>one</u> innovation point by protecting or restoring at least <u>75%</u> instead of <u>50% (or ½)</u> of the site area (<u>not including the building footprint</u>).

Project Phase: Schematic Design

<u>LEED Submittal</u> Phase: <u>Construction</u>

Related Code or Standard: None

Responsible Party: <u>Civil Engineer</u>, Owner and Contractor

SSc5.2: Site Development: Maximize Open Space (1 Point for NC and CS and Schools)

Purpose:
To promote biodiversity via a high ratio of open space to development footprint.

Credit Paths for NC and CS and Schools:
1) If local codes have open space requirements, provide open space with plantings in the project boundary to exceed the local zoning requirements by 25% and/or reduce the development footprint, i.e., the total area of parking, access road, hardscape and building footprint.

OR
2) Provide open space with planting area equal to the building footprint if there is no local zoning requirement. This requirement shall be maintained for the life of the building.

OR
3) Provide open space with planting area equal to 20% of the site in areas with a zoning ordinance but no open space (zero) requirements.

For all three paths:
a) If your project is already earning SSc2 and is located in an urban area, you can still use the vegetated roof areas to contribute to credit in this category.
b) If your project is already earning SSc2 and is located in an urban area, a pedestrian-oriented hardscape area can contribute to credit in this category. For this purpose, at least 25% of the open space counted shall be vegetated.
c) Vegetated, naturally designed ponds (ponds designed in naturalistic style) and wetlands with a side slope less than 1:4 can be counted as open space.

Submittals:
1) A site plan highlighting the qualifying open space
2) Calculations showing the qualifying open space meets or exceeds the credit requirements

Synergies:
Vegetated open space on-site may contribute to the following credits:
- SSc6.1: Storm Water Design: Quantity Control
- SSc6.2: Storm Water Design: Quality Control
- SSc7.1: Heat Island Effect: Non-Roof
- SSc7.2: Heat Island Effect: Roof

Possible Strategies and Technologies:
1) Do a site element survey and develop a master plan.
2) Locate the building on the site carefully and minimize its footprint to avoid disrupting the site.
3) Share facilities with neighbors, build multi-story buildings and place parking under the buildings to maximize open space.

Extra Credit (Exemplary Performance):
You can get <u>one</u> innovation point by <u>doubling</u> the requirements, i.e., <u>50%</u> instead of 25% for Credit Path #1 above, and <u>40%</u> instead of 20% for Credit Path #3 above.

Project Phase: Schematic Design

<u>LEED Submittal</u> Phase: <u>Design</u>

Related Code or Standard: None

Responsible Party: <u>Civil Engineer</u> and Contractor

SSc6.1: Storm Water Design: <u>Quantity</u> Control (<u>1</u> Point for NC and CS and Schools)

Purpose:
1) To increase on-site infiltration.
2) To reduce impervious cover.
3) To eliminate contaminants, and storm water run-off pollution, and
4) To limit disruption of natural water hydrology.

Credit Paths:
If existing imperviousness is <u>equal to or less than 50%</u>:
1) <u>Prevent</u> post-development peak quantity and discharge rate <u>from exceeding</u> that of pre-development for the one- and two-year, 24-hour design storms with the help of a storm water management plan.

OR
2) Apply a quantity control strategy, a stream channel protection strategy, and a storm water management plan to prevent excessive erosion from reaching receiving stream channels.

If existing imperviousness is <u>greater than 50%</u>:
3) <u>Decrease</u> the storm water runoff volume from the two-year, 24-hour design storms <u>by 25%</u> with the help of a storm water management plan.

Submittals:
1) Provide storm <u>quantities and rates</u> for pre- and post-development situations.
2) Provide a storm water plan <u>assessment</u> prepared by a civil engineer or other design professional.
3) Provide storm water management <u>strategies</u> and the <u>percentage of rainfall</u> that each is handling.

Synergies:
Reducing the quantity and rate of storm water runoff may contribute to the following credit:
- SSc6.1: Storm Water Design: <u>Quality</u> Control

Decreasing impervious surfaces may contribute to the following credits:
- SSc5.1: Site Development: Protect or Restore Habitat
- SSc5.2: Site Development: Maximize Open Space
- SSc7.1: Heat Island Effect: Non-Roof
- SSc7.2: Heat Island Effect: Roof

Harvesting rainwater may contribute to the following credits:
- WEc1: Water Efficient Landscaping
- WEc3: Water Use Reduction

Projects in densely developed areas may have difficulty providing space for storm water mitigation features, and have <u>negative</u> impact on SSc6.1. See "SSc2: Development Density and Community Connectivity."

Possible Strategies and Technologies:
1) Promote infiltration and maintain natural storm water flows.
2) Design pervious paving, vegetated roof, etc., to reduce impervious surface.
3) Reuse storm water for non-potable water uses like urinal and toilet flushing, landscape irrigation and custodial needs.
4) Guidelines for capturing and reusing storm water runoff: when and how it will be used? Drawdown, drainage area, conveyance system, pretreatment and pressurization.
5) Treatment ponds and underground facilities.

Extra Credit (Exemplary Performance): None

Project Phase: Schematic Design

<u>LEED Submittal</u> Phase: <u>Design</u>

Related Code or Standard: None

Responsible Party: <u>Civil Engineer</u>

SSc6.2: Storm Water Design: <u>Quality</u> Control (<u>1</u> Point for NC and CS and Schools)

Purpose:
1) To manage storm water runoff and
2) To limit pollution and disruption of natural water flows.

Credit Path for NC and CS and Schools:
1) Promote infiltration, reduce impervious paving and capture, and process storm water runoff from <u>90%</u> of the average annual rainfall with the help of a storm water management plan, and acceptable **BMPs (best management practices).**

AND

BMPs shall be able to remove <u>80%</u> of the average annual post-development **total suspended solids (TSS)** load per an existing monitoring report. BMP must meet one of the two criteria:

a) Comply with specifications and standards from a <u>local or state program</u> which has adopted these performance criteria;

b) Performance monitoring data existing in-field shows compliance with the standards. Data shall <u>comply with accepted protocol</u> like **TARP (Technology Acceptance Reciprocity Partnership),** Washington State Department of Ecology for BMP monitoring.

Submittals:
1) A <u>list</u> of **BMPs** to treat storm water and a description of <u>filtration contribution</u> of each, and the <u>percentage of rainfall treated</u> by each measure
2) For structural controls, a <u>list</u> and a description of the <u>pollutant removal</u> performance of each measure, and <u>percentage of rainfall treated</u> by each

Synergies:
BMP reduces runoff and may contribute to the following credit:
- SSc6.1: Storm Water Design: <u>Quantity</u> Control

Decreasing impervious surfaces may contribute to the following credits:
- SSc5.1: Site Development: Protect or Restore Habitat
- SSc5.2: Site Development: Maximize Open Space
- SSc7.1: Heat Island Effect: Non-Roof
- SSc7.2: Heat Island Effect: Roof

BMPs such as vegetated swales, rain gardens, and rainwater harvesting systems may contribute to the following credit:
- WEc1: Water-Efficient Landscaping

Possible Strategies and Technologies:
1) Promote infiltration and reduce imperviousness to reduced pollutant loading.
2) Use **nonstructural** techniques like <u>vegetated swales, raingardens, rainwater recycling, disconnecting imperviousness,</u> and
3) Use **structural** measures: <u>manhole treatment devices, rainwater cisterns and ponds.</u>
4) Use **alternative** surfaces like <u>pervious paving, vegetated roof, or grid paver,</u> etc.
5) Use <u>vegetated filters, open channels and wetlands</u> or other integrated **natural and mechanical treatment systems** to treat storm water runoff.
6) Apply <u>Environmentally Sensitive Design, Low Impact Development,</u> or other **sustainable design strategies**.

Note: there are three distinct climates in the United States. Depending on how much annual rainfall an area receives, it can be classified as an:
- **Arid watershed:** <u>20</u> inches or less rainfall annually
- **Semi-arid watershed:** <u>20 to 40</u> inches of rainfall annually
- **Humid watershed:** <u>40 inches or more</u> of rainfall annually.

For this credit, <u>treating runoff from 90% of the average annual rainfall equates to</u>:
a) Arid watersheds: <u>0.5</u> inches of rainfall.

b) Semi-arid watersheds: <u>0.75</u> inches of rainfall.
c) Humid watersheds: <u>1</u> inch of rainfall.

Extra Credit (Exemplary Performance): None

Project Phase: Schematic Design

<u>LEED Submittal</u> Phase: <u>Design</u>

Related Code or Standard:
1) <u>BMP</u>: Best Management Practice
2) <u>EPA</u>: Environmental Protection Agency

Responsible Party: <u>Civil Engineer</u>

SSc7.1: Heat Island Effect: Non-Roof (<u>1</u> Point for NC and CS and Schools)

Purpose:
To minimize a heat island's impact on wildlife habitat, human beings and microclimate. **Heat islands** refer to the extra thermal gradient in developed areas when compared with undeveloped areas.

Credit Paths for NC and CS and Schools:
1) For <u>50%</u> of the parking lots, roads, sidewalks, courtyards, and other site hardscape, use <u>any combination of</u>
   a) <u>Open grid pavement</u> system (50% pervious minimum).
   b) Paving materials with a SRI (Solar Reflectance Index) of <u>29 or more</u>.
   c) <u>Shade</u> from a tree canopy (within <u>five</u> years of occupancy). Trees shall be planted at the time of occupancy.
   d) <u>Shade</u> from a structure covered by <u>qualified solar panels</u>. **Qualified solar panels** produce energy to offset a portion of the nonrenewable resource use.
   e) <u>Shade</u> from a structure or architectural device with a SRI of <u>29 or more</u>.

   OR
2) Provide cover/roof for <u>50%</u> or more of the parking spaces, the cover/roof shall have a SRI of <u>29 or more</u>, OR be covered by <u>qualified solar panels</u>, OR be a <u>vegetated</u> green roof. The qualified parking space undercover can be under buildings, decks, roofs or underground.

Submittals:
1) **For shaded condition:** A <u>site plan</u> highlighting all nonroof hardscape. Label <u>hardscape</u> used for the credit and list the <u>SRI</u> for the compliant surface.
2) **For covered parking condition:** Show the number of <u>spaces</u> and portion <u>covered</u>, and the <u>SRI</u> <u>for roofs</u> covering the parking areas.

Synergies:
Locating the parking structure underground may contribute to the following credit:

- SSc5.2: Site Development: Maximize Open Space

Open-grid pavement may contribute to the following credits:

- SSc6.1: Storm Water Design: <u>Quantity</u> Control
- SSc6.2: Storm Water Design: <u>Quality</u> Control

If you are using vegetation to shade hardscape, you may want to coordinate with the following credit:

- WEc1: Water-Efficient Landscaping

Possible Strategies and Technologies:
1) Use high-reflectance materials for hardscape and provide shaded surfaces with landscaped features.
2) Use open grid paving, vegetated roofs or high-albedo (high-reflectivity) materials to replace conventional surfaces like sidewalks, roads and roofs, etc., to reduce heat absorption.

SRI (Solar Reflectance Index) refers to a built surface's capacity to reflect solar heat in terms of a small rise in temperature. SRI for a standard <u>white surface (emittance 0.90, reflectance 0.80) is 100</u>, SRI for a standard <u>black surface (emittance 0.90, reflectance 0.05) is 0</u>. You can use a material's emittance value and reflectance value to calculate its SRI. **SRI** is calculated per <u>ASTM</u> E 1980-01. Emittance is calculated per ASTM C 1371 or ASTM E 408. **Reflectance** is calculated per ASTM C1549, ASTM E903 or ASTM E1918. You can find default SRI values in the related USGBC reference guide. You can also find product information at www.coolroofs.org (Cool Roof Rating Council website).

Extra Credit (Exemplary Performance):
You can get one innovation point by doubling the requirements, i.e., 100% instead of 50% for both credit path #1 and #2 above.

Project Phase: Schematic Design

<u>LEED Submittal</u> Phase: <u>Construction</u>

Related Code or Standard: None

Responsible Party: <u>LEED AP, Landscape Architect, Civil Engineer</u>, and Contractor

SSc7.2: Heat Island Effect: Roof (<u>1</u> Point for NC, CS, and Schools)

Purpose:
To minimize heat islands' impact on wildlife habitat, human beings and microclimate. **Heat island** refers to the extra thermal gradient in developed areas when compared to undeveloped areas.

Credit Paths for NC, CS, and Schools:
1) For at least <u>75%</u> of the roof surface, use roofing materials with a SRI of <u>78</u> or more for low-sloped roof (Slope 2:12 or less) and a SRI of <u>29</u> or more for steep-sloped roof (Slope <u>more than 2:12</u>).

Roofing materials with a SRI lower than the ones listed above may be used if they meet the following:

$$\frac{SRI\,of\,Installed\,Roof}{Re\,quired\,SRI} \times \frac{Area\,of\,Roof\,Meeting\,Minimum\,SRI}{Total\,Roof\,Area} \geq 75\%$$

OR

2) Use a vegetated roof for <u>50% or more</u> of the roof area.

OR

3) Use a combined vegetated roof and high <u>albedo (reflectivity)</u> surfaces that meet the following:

$$\frac{Area\,of\,Vegetated\quad Roof}{0.5} + \frac{Area\,of\,SRI\,Roof}{0.75} \geq Total\quad Roof\quad Area$$

AND

Use roof materials with a SRI of <u>78</u> or more for low-sloped roof (Slope <u>2:12 or less</u>) and a SRI of <u>29</u> or more for steep-sloped roof (Slope <u>more than 2:12</u>).

Submittals:
1) A <u>roof plan</u> showing the total roof area and the vegetated roof area and/or the highly reflective roof area
2) A <u>list</u> of roofing products used on the project and their SRI, slopes, reflectance percentage, emittance percentage, and product <u>specifications</u> confirming the information.

Synergies:
Vegetated roofs may contribute to the following credits:
- SSc5.1: Site Development: Protect or Restore Habitat
- SSc5.2: Site Development: Maximize Open Space
- SSc6: Storm Water Design

Vegetated roofs also reduce the harvested rainwater; they may also have a <u>negative</u> impact on the following credit:
- WEc3: Water Use Reduction

Vegetated roofs and/or highly reflective roofing may reduce cooling loads and contribute to the following credit:
- EAc1: Optimize Energy Performance

Possible Strategies and Technologies:
Use a vegetated roof and a high albedo roof to reduce heat absorption.

Extra Credit (Exemplary Performance):
You can get one innovation point if 100% of the roof (excluding skylights, photovoltaic panel and HVAC equipment) is made of a green roof system.

Project Phase: Schematic Design

LEED Submittal Phase: <u>Design</u>

Related Code or Standard: <u>ASTM</u>

Responsible Party: <u>LEED AP</u> and Contractor

SSc8: Light Pollution Reduction
(<u>1</u> Point for NC, CS, and Schools)

Purpose:
1) To reduce development impacts on nocturnal environments.
2) To reduce glare, improve nighttime visibility, and reduce sky-glow and light trespass from the site and building to increase night sky access.

Credit Paths for NC, CS, and Schools:
1) For <u>Exterior</u> Lighting
Light only the areas needed for comfort and safety. Lighting power density shall not exceed ANSI/<u>ASHRAE/IESNA Standard 90.1-2007 (with addenda)</u>, for the classified zone (See various zone definitions below). The project team must provide justification for the selected lighting zone. Exterior lighting controls shall meet section 9.4.1.3 of ANSI/<u>ASHRAE/IESNA Standard 90.1-2007 (without amendments)</u>.

AND
2) For <u>Interior</u> Lighting
Use <u>automatic controls</u> to turn off 50% of non-emergency interior lighting with a direct line of sight to any openings in the building envelope between 11 p.m. to 5 a.m. Provide occupant sensing devices or capacity for <u>manual override</u> during after-hours use. The override shall not last more than 30 minutes.

OR
All openings in the building envelope that have a direct line of sight to any non-emergency interior lighting should have shielding (closed/controlled by automatic device between 11:00 p.m. and 5:00 a.m.) to allow for less than 10% transmittance.

Note: This credit does <u>not</u> apply to <u>warehouses, manufactured homes, three-story buildings, or single family houses</u>.

Per <u>IESNA RP-33</u>, all projects shall be classified in <u>one</u> of the following zones and shall comply with the related requirements:

LZ1-Dark (developed areas within <u>rural</u> areas, national parks, and state parks forest land. The population density is <u>less than 200</u> people per square mile).
You should design building exterior lights and site lights to produce an initial illuminance value no higher than <u>0.01</u> vertical and horizontal foot candles at the site boundary and beyond. You should document that <u>0%</u> of the total initially designed fixture lumens are emitted at an angle of 90 degree or higher from <u>nadir (straight down)</u>.

LZ2-Low (light industry with limited nighttime use, neighborhood business district, primarily residential zones and residential mixed-use areas. The population density is between 200 and 3,000 people per square mile).
You should design building and site mounted lights and exterior lights to produce an initial illuminance value no higher than 0.10 vertical and horizontal foot candles at the site boundary and no higher than 0.01 horizontal foot candles 10 feet beyond the site boundary. You should document that 2% or less of the total initially designed fixture lumens are emitted at an angle of 90 degree or higher from nadir (straight down).

LZ3-Medium (All areas not included in LZ1, LZ2 or LZ3, like commercial/industrial, and high-density residential. The population density is more than 3,000 people per square mile).
You should design building and site mounted lights and exterior lights to produce an initial illuminance value no higher than 0.20 vertical and horizontal foot candle at the site boundary and no higher than 0.01 horizontal foot candles 15 feet beyond the site boundary. You should document that 5% or less of the total initially designed fixture lumens are emitted at an angle of 90 degree or higher from nadir (straight down).

LZ4-High (high activity commercial districts in major metropolitan areas, like major city centers and entertainment districts. The population density is more than 100,000 people per square mile).
You should design building and site mounted lights and exterior lights to produce an initial illuminance value no higher than 0.60 vertical and horizontal foot candle at the site boundary and no higher than 0.01 horizontal foot candles 15 feet beyond the site boundary. You should document that 10% or less of the total initially designed fixture lumens are emitted at an angle of 90 degree or higher from nadir (straight down).

For **LZ2, LZ3** and **LZ4**, if your project boundary is next to a public right-of-way, you can use the curb line instead of the site boundary for light trespass requirements.

For **all zones**, for a single luminaire and the related illuminance located at the intersection of a public roadway accessing the site and a private driveway, you can use the centerline of the public roadway as the site boundary for a length of twice the driveway width centered at the centerline of the driveway.

Credit Paths for Schools:
Sports Field Lighting (Physical Education Spaces)
Per ANSI/ASHRAE/IESNA Standard 90.1-2007 section 9.4.5, playing fields and similar physical education spaces do not need to comply with the lighting density requirements of this credit, except that all sports lighting shall be turned off automatically by 11 p.m., and a manual override shall be provided to prevent interruption of sports events.

Trespass Calculations
Must be done for two conditions:
   a) With the sports lighting off and all other site lighting on, the criteria above shall be met.
   b) With only the sports lighting on, the following criteria shall be met:
      • **LZ1** = 0.10fc (footcandles) at the site boundary, and down to 0.01fc within 10 feet of the boundary
      • **LZ2** = 0.30fc at the site boundary, and down to 0.01fc within 10 feet of the boundary
      • **LZ3** = 0.80fc at the site boundary, and down to 0.01fc within 15 feet of the boundary
      • **LZ4** = 1.50fc at the site boundary, and down to 0.01fc within 15 feet of the boundary

Submittals:

For Interior Lighting:
1) <u>Drawings</u> showing the locations of automatic controls, sequence of operation, and specifications or the building operation plan
2) <u>Drawings</u> and specifications for automatic shading devices, or <u>cutsheets</u> (product data) showing that they block <u>90%</u> of the light, sequence of operation, and specifications or the building operation plan

For Exterior Lighting:
1) <u>Zone classification</u> of the project site
2) Product <u>data</u> for lamps used
3) A description of the <u>light trespass analysis procedure</u> conducted
4) A <u>site photometric plan</u> for the parking area showing footcandle summaries and lighting ratio

For Schools:
1) A <u>site photometric plan</u> for the sports field showing compliance with lighting level limits.
2) <u>Drawings</u> showing the locations of automatic controls, sequence of operation, and specifications or the building operation plan

Synergies:
Energy savings beyond the baseline per the lighting densities of ASHRAE 90.1-2007, and savings from integrated automatic controls may contribute to the following credits:
- EAc1: Optimize Energy Performance
- IEQc6.1: Controllability of Systems: Lighting

Possible Strategies and Technologies:
1) Set up site lighting criteria to avoid night sky pollution and off-site lighting, but still maintain safe light levels.
2) Use a computer to model the site lighting and minimize site lighting when possible.
3) Use low-angle spotlights, low-reflectance surfaces and full cutoff luminaires to reduce light pollution.

Extra Credit (Exemplary Performance): None

Project Phase: Schematic Design

LEED Submittal Phase: <u>Design</u>

Related Code or Standard:
1) <u>ASHRAE/IESNA Standard 90.1-2007 (with errata but without addenda)</u>
2) <u>ANSI</u>
3) <u>California Energy Code Title 24</u>

Responsible Party: <u>Lighting Designer and LEED AP</u>

SSc9: Tenant Design and Construction Guidelines (<u>1</u> Point for CS)

Note: Detailed discussions have been omitted since this credit is for CS only, and is unlikely to be tested on the LEED Green Associate Exam.

SSc9: Site Master Plan (<u>1</u> Point for Schools)

Note: Detailed discussions have been omitted since this credit is for schools only, and is unlikely to be tested on the LEED Green Associate Exam.

SSc10: Joint Use of Facilities (<u>1</u> Point for Schools)

Note: Detailed discussions have been omitted since this credit is for schools only, and is unlikely to be tested on the LEED Green Associate Exam.

SS Summary and Mnemonics:

SS Credit Name	Extra Credit	Responsible Party
SSp1 (SS Prerequisite 1): <u>C</u>onstruction Activity Pollution Prevention (<u>Required</u> for NC and CS and Schools)	0	Civil Engineer and Contractor
*SSp2: Environmental Site Assessment (<u>Required</u> for Schools Only, N/A for NC & CS)	0	Soils Engineer and Owner
SSc1: <u>S</u>ite Selection (<u>1</u> Point for NC and CS and Schools)	0	Civil Engineer and Owner
SSc2: Development <u>D</u>ensity and Community Connectivity (<u>5</u> Points for NC and CS, <u>4</u> Points for Schools)	1 extra point for <u>doubling</u> the requirements, i.e. <u>120,000 s.f per acre</u>	LEED AP and Owner
SSc3: <u>B</u>rownfield Redevelopment (<u>1</u> Point for NC and CS and Schools)	0	Civil Engineer and Owner
SSc4.1: <u>A</u>lternative Transportation: Public Transportation Access (<u>6</u> Points for NC and CS, <u>4</u> points for Schools)	1 extra point for <u>doubling</u> the requirements, i.e., <u>2</u> train stations or <u>4</u> bus lines.	LEED AP and Owner
SS Credit 4.2: Alternative Transportation: Bicycle Storage and Changing Rooms (<u>1</u> Point for NC and Schools, <u>2</u> points for CS)	1 extra point for credit 4 for program that results in <u>quantifiable</u> automobile use reduction.	Architect and LEED AP

SSc4.3: Alternative Transportation: Low Emitting and Fuel Efficient Vehicles (3 Point for NC and CS, 2 points for Schools)	Same as above	Architect, MEP Engineer, LEED AP, and Owner
SSc4.4: Alternative Transportation: Parking Capacity (2 points for NC and CS and Schools)	Same as above	Civil Engineer and Owner
SSc5.1: Site Development: Protect or Restore Habitat (1 Point for NC and CS and Schools)	1 extra point for 75% instead of 50%.	Civil Engineer, Owner, and Contractor
SSc5.2: Site Development: Maximize Open Space (1 Point for NC and CS and Schools)	1 innovation point by doubling the requirements, i.e., 50% instead of 25% for Credit Path #1, and 40% instead of 20% for Credit Path #3.	Civil Engineer and Contractor
SSc6.1: Storm Water Design: Quantity Control (1 Point for NC and CS and Schools)	0	Civil Engineer
SSc6.2: Storm Water Design: Quality Control (1 Point for NC and CS and Schools)	0	Civil Engineer
SSc7.1: Heat Island Effect: Non-Roof (1 Point for NC and CS and Schools)	1 innovation point by doubling the requirements, i.e., 100% instead of 50% for both credit path #1 and #2.	LEED AP, Landscape Architect, Civil Engineer, and Contractor
SSc7.2: Heat Island Effect: Roof (1 Point for NC and CS and Schools)	1 extra point if 100% of the roof (excluding sky-lights, photovoltaic panel and HVAC equipment) is made of a green roof system.	LEED AP and Contractor
SSc8: Light Pollution Reduction (1 Point for NC and CS and Schools)	0	Lighting Designer and LEED AP
*SSc9: Tenant Design and Construction Guidelines (1 Point for CS, N/A for NC & Schools)	0	Architect and Owner
*SSc9: Site Master Plan (1 Point for Schools, N/A for NC & CS)	0	Master Planner or Architect
*SSc10: Joint Use of Facilities (1 Point for Schools, N/A for NC & CS)	1 extra ID point for meeting the requirements of 2 of the 3 credit paths.	Owner (School Administrator)

Mnemonics: Carole Smith Does Business As "So Sweet, Hot Love." (See bold and underlined letters in the credit names on table above also). Of course, you can create your own **mnemonics** to help you memorize the credits above as well.

Note: * indicates prerequisite or credit NOT applicable to all LEED rating systems. Detailed discussions have been omitted since the information is unlikely to be tested on the LEED Green Associate Exam.

B. Water Efficiency (WE)

Overall purpose:
1) Water Efficient Landscaping (Outdoor Water): Reduction of 50% from a calculated midsummer baseline; no potable water use, or no irrigation
2) Innovative Wastewater Technologies (Indoor Water): Reuse water when legal, safe and appropriate
3) Water Use Reduction (Indoor Water)
   Mnemonics:
   Love In Universe, or LIU (See underlined letters above also).

Core concepts:
1) Regulation of indoor water
   - Save as much indoor potable water as possible
   - Use water efficiently
2) Outdoor water
   - Save as much outdoor potable water as possible
   - Use water efficiently
3) Process water
   - Reduce the need for potable water when supplying process water
   - Use water efficiently

Recognition, regulation and incentives:
1) Recognition
   - WaterSense product label sponsored by the EPA
2) Regulation (Requirements and Goals)
   - Mandatory federal water efficiency requirements per the Energy Policy Act (EPAct) of 2005 and Executive Order 13423 (2007): All federal facilities shall reduce water use intensity by 2% per year between 2006 and 2015 to reach a total reduction of 20%
   - Energy Policy Act (1992): Mandatory requirements requiring the use of water conserving plumbing fixtures in industry, commercial and residential buildings.
3) Financial incentives
   Local/state rebates and credits for water saving devices

Overall strategies and technologies:
Note: Not **all** strategies and technologies have to be used simultaneously for your project development.
1) **Reduce indoor potable water demand**
   Use non-potable water: reuse graywater, and capture and use rainwater; reduce water use via innovative wastewater treatment; use water efficiently: use waterless or high efficiency fixtures
2) **Reduce outdoor potable water demand**
   Use native and/or adapted and drought-tolerant plants; use non-potable water: capture and use storm water for landscape irrigation; use water efficiently: use drip-irrigation or other high efficiency technologies.
3) **Reduce potable water need for process water**
   Use process water efficiently: use of controls and sensors; efficient management of cooling tower water; use non-potable water.

Specific Technical Information:

WEp1: Water Use Reduction (<u>Indoor</u> Water, <u>Required</u> for NC, CS, and Schools)

Purpose:
To reduce the pressure on wastewater systems and municipal water supplies by saving as much water as possible within buildings.

Note: This prerequisite is for reducing <u>indoor</u> water use. Some questions on practice exams or real LEED exams may test your understanding of whether a prerequisite or a credit is for <u>indoor</u> water or for <u>outdoor</u> water. I have read these kinds of sample questions on LEED exams.

Credit Path for NC, CS, and Schools:
You should meet the <u>Energy Policy Act of 1992</u> fixture performance requirements, and then apply strategies to save <u>20%</u> of water when compared with the water use baseline calculation for the building (<u>not</u> including landscape <u>irrigation</u>). Your calculation should only include <u>urinals, water closets, showers, pre-rinse spray valves, kitchen sink faucets, lavatory faucets</u>, and shall be based on estimated occupant use.

Some buildings have several shifts, and may include the following:
1) Residents
2) Full-Time staff
3) Part-Time staff
4) Transients (retail customers, student visitors, etc.)

You can identify the number of occupants by **occupancy type**, i.e., two people for a one-bedroom unit, and three people for a two-bedroom unit, etc. If you do not know the occupancy (like CS projects or mixed-use projects in the early design stages where you do not know who will be the tenants), then you can use the "Default Occupancy Factors" at the appendixes.

FTE = number of hours of occupancy/<u>8</u>

You can estimate the number of transients for your projects. Use a daily <u>average</u> number over a one year period.

Use your best judgment to decide if someone should be reported as a FTE or a transient. For example, a volunteer who works at the school 4 hours each day can be considered a FTE with a value of 0.5, and an individual who attends a basketball game can be reported as a visitor.

Use a <u>1 to 1</u> male to female ratio for your projects unless they have a specific ratio. You need to describe your special ratio with a narrative.

Important data to memorize (The numbers in the two tables below are <u>very</u> important. Almost every LEED exam tests them)

Fixture uses per day:

Non-residential Projects

	Water Closet	Urinal	Lavatory	Shower	Kitchen Sink
FTE (including Student FTE)					
Female	3	0	3	0.1 0 for student FTE	1
Male	1	2	3	0.1 0 for student FTE	1
Transients (Student Visitors)					
Female	0.5	0	0.5	0	0
Male	0.1	0.4	0.5	0	0
Transients (Retail Customers)					
Female	0.2	0	0.2	0	0
Male	0.1	0.1	0.2	0	0

Residential Projects

	Water Closet	Urinal	Lavatory	Shower	Kitchen Sink
Female	5	N/A	5	1	4
Male	5	N/A	5	1	4

Note:
1) Lavatory faucets are counted as a 60 second duration per use for **residential** projects, a 15 second duration for **non-residential** projects, and a 12 second duration when equipped with autocontrol.
2) Kitchen sink faucets are counted as a 60 second duration per use for **residential** projects, and a 15 second duration for **non-residential** projects
3) Showers are counted as a 480 second duration for **residential** projects, and 300 second duration for **non-residential** projects.
4) This table applies to NC, CS, schools, commercial and residential projects. I have combined several tables into one table to save you time.

Flow Rates (GPF or GPM or GPC):

	Water Closet	Urinal	Lavatory	Shower	Kitchen Sink
Conventional (baseline)	1.6 gpf except blow-out fixture at 3.5 gpf	1.0	**Residential** 2.2gpm at 60psi **Commercial** 2.2gpm at 60 psi for private app. 0.5gpm at 60 psi for public use 0.25gpc (gallons per cycle) for metering faucets	2.5 gpm at 80psi per shower stall	2.2gpm at 60 psi
EPA Water-Sense Standard or High-efficiency	1.28 gpf	0.5 gpf	1.5 gpm for private lavatory faucets and aerators	≤2.0 gpm	
HET, single flush pressure assist	1 gpf	0.5 gpf for HEU			
HET, dual flush (full-flush)	1.6 gpf				
HET, dual flush (low-flush)	1.1 gpf				
HET, foam flush	0.05 gpf				
Low-Flow	1.1 gpf	0.5 gpf	1.8 gpm	1.8 gpm	≤2.2 gpm
Ultra Low-Flow	0.8 gpm		0.5 gpm		
Non-Water Urinal or Composting Toilet		0			

Note:

1) Private app. (application) means hospital patient rooms, motel and hotel rooms, etc.
2) Commercial **pre-rinse spray valves** for food service application ≤ **1.6gpm**
3) This table applies to **non-residential** (NC, CS, school, commercial, etc.) projects and **residential** projects. I have combined several tables into one table to save you time.
4) This table is based on the **Energy Policy Act of 1992**, **EPA WaterSense** Standard, Uniform Plumbing Code (**UPC**) and International Plumbing Code (**IPC**) standards for plumbing fixture water use. **WaterSense** is a partnership program sponsored by the **EPA.** The **EPA WaterSense** standard exceeds the IPC and UPC requirements in some cases.
5) The following fixtures shall <u>not</u> be included in the water use calculations for this credit, but they may be included for extra ID points for WRc3, Water Use Reduction:
   Automatic commercial ice makers, commercial dishwashers, commercial steam cooker, residential clothes washers, commercial (family-size) clothes washers, standard and compact residential dishwashers.
6) High-efficiency (HE) fixtures include: single-flush gravity fed, high-efficiency toilets (HET), high-efficiency urinals (HEU), etc.

   Note: HETs are available in different flush types: single-flush gravity fed (like a conventional toilet), single flush pressure assist, and dual-flush in both gravity fed and pressure assist.

Calculating Annual Occupancy for Schools with Multiple Sessions

Formula 1:
Session Percentage = Number of Days in Session/Annual Days of Operation

Formula 2:
Annual Occupants by Gender = (Session A Percentage × Session A FTE by Gender) + (Session B Percentage × Session B FTE by Gender)

Formula 3:
Average Daily Use per Person = (Session A Percentage × Session A Daily Use per Person) + (Session B Percentage × Session B Daily Use per Person)

Submittals:

1) Number and type of <u>occupants</u>
2) <u>Manufacturer's data</u> (cutsheets) showing manufacturer, model, and water consumption rates of each fitting and fixture
3) A definition of the <u>usage group</u>
4) A <u>list</u> of plumbing fixtures by usage group

Synergies:

Increasing the use of rainwater, graywater, and reducing the demand on local water aquifers may contribute to the following credits:

- SSc6.1: Storm Water Design: <u>Quantity</u> Control
- SSc6.2: Storm Water Design: <u>Quality</u> Control
- WEc1: Water-Efficient Landscaping
- WEc2: Innovative Wastewater Technologies
- WEc3: Water Use Reduction
- WEc4: Process Water Use Reduction

Possible Strategies and Technologies:

1) Use graywater or storm water for custodial uses, flushing urinals and toilets, and other non-potable use.
2) Use water-saving and waterless plumbing fixtures like waterless urinals and com-posting toilet systems, and occupant sensors to reduce wastewater volumes.

Extra Credit (Exemplary Performance): None

Project Phase: Schematic Design

LEED Submittal Phase: Design

Related Code or Standard:

1) Energy Policy Act (EPAct) of 1992 (and as amended)
2) Energy Policy Act (EPAct) of 2005
3) International Associations of Plumbing and Mechanical Officials, Publication IAPMO/ American National Standards Institute UPC 1-2006, Uniform Plumbing Code 2006, Section 402.0, Water-Conserving Fixtures and Fittings
4) International Code Council, International Plumbing Code 2006, Section 604, Design of Building Water Distribution System

Responsible Party: MEP Engineer

WEc1: Water-Efficient Landscaping (Outdoor Water, 2 to 4 Points for NC, CS, and Schools)

Purpose:
To reduce the use of subsurface or surface natural water or potable water available on or near the job site for landscape irrigation.

Credit Path for NC, CS, and Schools:

1) 50% reduction (2 points)
Based on calculated mid-summer (July) baseline case, reduce 50% of the potable consumed by irrigation.

You can achieve this water reduction from any combination of the following:
   a. Capture and use rainwater.
   b. Recycle wastewater.
   c. For non-potable irrigation use, use water treated and delivered by a public agency.
   d. Efficient irrigation.
   e. Choose the right plant species.
   f. Microclimate and density factor.
   g. You can pump groundwater seepage away from the vicinity of building foundations and

slabs and use it for irrigation, but you must demonstrate doing so does <u>not</u> affect site storm water management systems.

2) No Potable Water Use or No Irrigation (4 points)

Meet the requirements for Path One.

AND
a. Use plants that do <u>not</u> need a permanent irrigation system. You can use temporary irrigation systems for plant establishment only if you <u>remove them within 18 months of installation</u>.

OR
b. <u>Do not use any potable water</u>. For non-potable irrigation use, you can only use <u>recycled graywater, recycled wastewater, captured rainwater or water treated and delivered</u> by a public agency.

Submittals:
1) <u>Calculations</u> of baseline and design cases to show <u>percentage</u> of water use reduction
2) <u>Description</u> of which portion of the landscape irrigation will come from <u>nonpotable sources</u>
3) A landscape <u>plan</u> showing an <u>irrigation system</u> and a <u>planting schedule</u>

Synergies:
Using adapted or native <u>vegetation</u> may contribute to the following credits:
- SSc5.1: Site Development: Protect or Restore <u>Habitat</u>
- SSc5.2: Site Development: Maximize <u>Open Space</u>
- SSc7.2: <u>Heat Island Effect</u>: Roof

<u>Rainwater capture systems</u> can reduce potable water use and may contribute to the following credits:
- SSc6.1: Storm Water Design: <u>Quantity</u> Control
- SSc6.2: Storm Water Design: <u>Quality</u> Control

<u>Landscaping</u> can provide shade for south-facing windows and hardscape, mitigate climate conditions, reduce building energy use, and may contribute to the following credits:
- SSc7.1: Heat Island Effect: <u>Non-Roof</u>
- EAp2: <u>Minimum</u> Energy Performance
- EAc1: <u>Optimize</u> Energy Performance

Possible Strategies and Technologies:
1) Do a climate and/or soils analysis and choose native or adapted plants to reduce irrigation water use.
2) Use a climate-based controller and/or high efficiency equipment, like drip irrigation system, or micro-irrigation system.
3) Use <u>mulching, composting and alternative mowing</u> to minimize turf area.

Extra Credit (Exemplary Performance): None

Project Phase: Schematic Design

LEED Submittal Phase: <u>Design</u>

Related Code or Standard: None

Responsible Party: <u>Landscape Architect</u>

WEc2: Innovative Wastewater Technologies (<u>Indoor</u> Water, <u>2</u> Points for NC, CS, and Schools)

Purpose:
To reduce demand for potable water and generate less wastewater, also increase the amount of water recharged to local aquifers.

Credit Paths for NC, CS, and Schools:
1) Use <u>non-potable water</u> (recycled graywater, captured rainwater, and municipally treated wastewater) or <u>water-saving plumbing fixtures</u> (urinals, toilets) to reduce <u>50%</u> of the potable water use for building sewage conveyance.

   OR
2) Follow tertiary standards and treat <u>50%</u> of the on-site wastewater. You can use a <u>living machine/natural system</u> or a <u>packaged mechanical wastewater treatment system</u>. The treated water shall be used on-site or infiltrated.

Submittals:
1) Calculate number and type of <u>occupants</u>
2) <u>Manufacturer's data</u> showing manufacturer, model, water consumption rate of each fitting and fixture
3) Information on <u>system schematics</u> and <u>capacity</u> of graywater and rainwater systems

Synergies:
Reusing graywater, harvesting rainwater, and reducing demand on local water aquifers may <u>contribute</u> to the following credits:
- SSc6.1: Storm Water Design: <u>Quantity</u> Control
- SSc6.2: Storm Water Design: <u>Quality</u> Control
- WEp1: Water Use Reduction
- WEc1: Water-Efficient Landscaping
- WEc3: Water Use Reduction
- WEc4: Process Water Use Reduction

On the other hand, <u>active</u> systems for on-site wastewater treatment or reuse may need extra energy, and may <u>affect</u> the following credits:
- EAp1: <u>Fundamental</u> Commissioning of the Building Energy Systems
- EAc3: <u>Enhanced</u> Commissioning
- EAc5: <u>Measurement and Verification</u>

Possible Strategies and Technologies:
1) <u>Reuse graywater or storm water</u> for mechanical and/or natural on-site waste treatment system, or sewage conveyance.
2) Use <u>water-saving and waterless plumbing fixtures</u> like waterless urinals and composting toilet systems to reduce wastewater volumes.
3) Use <u>high-efficiency filtration systems</u>, construct wetlands, or use a packaged biological nutrient removal system.

Extra Credit (Exemplary Performance):
You can get one innovation point by doubling the requirements, i.e., reduce or treat and reuse <u>100%</u>.

Project Phase: Schematic Design

LEED Submittal Phase: <u>Design</u>

Related Code or Standard:
1) <u>Energy Policy Act (EPAct) of 1992 (and as amended)</u>
2) <u>Energy Policy Act (EPAct) of 2005</u>
3) <u>International Associations of Plumbing and Mechanical Officials, Publication IAPMO/ American National Standards Institute UPC 1-2006, Uniform Plumbing Code 2006, Section 402.0, Water-Conserving Fixtures and Fittings</u>
4) <u>International Code Council, International Plumbing Code 2006, Section 604, Design of Building Water Distribution System</u>

Responsible Party: <u>MEP Engineer</u>

WEc3: Water Use Reduction (<u>Indoor</u> Water, <u>2</u> to <u>4</u> Points for NC and CS and Schools)

Purpose:
To reduce pressure on wastewater systems and municipal water supplies by saving as much water <u>within</u> buildings as possible.

Credit Path:
You should meet the <u>Energy Policy Act of 1992</u> fixture performance requirements, and then apply strategies to save water when compared with the water use baseline calculation for the building. Your calculation should only include <u>urinals, water closets, showers, pre-rinse spray valves, kitchen sink faucets, and lavatory faucets</u> and shall be based on estimated occupant use.

Percentage of Water Use Reduction	Points
30%	2
35%	3
40%	4

Note: This percentage of water use reduction is based on the water use baseline. See Tables and notes in **WEp1** for water use baseline information.

Submittals: See WEp1

Synergies: See WEp1

Possible Strategies and Technologies:
1) Use <u>graywater or storm water</u> for custodial uses, flushing urinals and toilets, and other non-potable use.
2) Use <u>water-saving and waterless plumbing fixtures</u> like waterless urinals and composting toilet systems, and occupant sensors to reduce wastewater volumes.

Extra Credit (Exemplary Performance):
You can get one innovation point by saving <u>45%</u> of <u>potable</u> water.

Project Phase: Schematic Design

<u>LEED Submittal</u> Phase: <u>Design</u>

Related Code or Standard:
<u>Energy Policy Act of 1992</u>

Responsible Party: <u>MEP Engineer</u>

WEc4: <u>Process</u> Water Use Reduction (<u>1</u> Point for Schools ONLY)

Note: Detailed discussions have been omitted since this credit is for schools only, and is unlikely to be tested on the LEED Green Associate Exam.

WE Summary and Mnemonics:

WE Credit Name	Extra Credit	Responsible Party
WEp1: Water <u>U</u>se Reduction (<u>Indoor</u> Water, <u>Required</u> for NC and CS and Schools)	0	<u>MEP Engineer</u>
WEc1: Water Efficient <u>L</u>andscaping (<u>Outdoor</u> Water, <u>2</u> to <u>4</u> Points for NC and CS and Schools)	0	<u>Landscape Architect</u>
WEc2: <u>I</u>nnovative Wastewater Technologies (<u>Indoor</u> Water, <u>2</u> Points for NC and CS and Schools)	1 extra ID point by doubling the requirements, i.e., reduce, treat, or reuse <u>100%.</u>	<u>MEP Engineer</u>
WEc3: Water <u>U</u>se Reduction (<u>Indoor</u> Water**,** <u>2</u> to <u>4</u> Points for NC and CS and Schools)	1 extra ID point by saving <u>45%</u> of water.	<u>MEP Engineer</u>
*WEc4: <u>Process</u> Water Use Reduction (<u>1</u> Point for Schools ONLY)	1 extra ID point for saving <u>40%</u> of <u>process</u> water.	<u>MEP Engineer</u>

Mnemonics: <u>U</u>niversal <u>L</u>ifeline <u>I</u>s <u>U</u>niversal for the <u>P</u>oor (See bold and underlined letters in credit names on table above also). Of course, you can create your own **mnemonics** to help you memorize them.

Note: * indicates prerequisite or credit NOT applicable to all LEED rating systems. Detailed discussions have been omitted since the information is unlikely to be tested on the LEED Green Associate Exam.

C. Energy and Atmosphere (EA)

Overall purpose:
1) <u>S</u>ave energy
2) <u>P</u>romote the supply of renewable energy
3) <u>F</u>undamental Commissioning of the Building Energy Systems
4) <u>M</u>inimum Energy Performance
5) <u>F</u>undamental Refrigerant Management
6) <u>O</u>ptimize Energy Performance
7) <u>O</u>n-Site Renewable Energy
8) <u>E</u>nhanced Commissioning
9) <u>E</u>nhanced Refrigerant Management
10) <u>M</u>easurement and Verification
11) <u>G</u>reen Power

Mnemonics:
<u>So</u>, <u>P</u>lease <u>F</u>orward <u>M</u>y <u>F</u>iles!
<u>Oh</u>! <u>O</u>prah <u>E</u>nters <u>E</u>nergy <u>M</u>anagement <u>G</u>roup. (See underlined letters above also).

Core concepts:
1) Energy Efficiency and Demand
   - Understand energy criteria
   - Save energy
   - Measure the performance of energy
2) The Supply of Energy
   - Buy off-site renewable energy
   - Generate on-site renewable energy

Recognition, regulation and incentives:
1) **Recognition:** Energy Star Program, Target Finder Rating Tool, etc.
2) **Regulation (Requirements and Goals):** See detailed discussion under each credit.
3) **Incentives**
   - Private Sector: lower risk and lower premiums for property insurance, social responsibility of corporations, and availability of money
   - Public Sector: tax rebates and credits, incentive for development, expedited plan review and permit processing, and technology-based measures.

Overall strategies and technologies:

Note: Not **all** strategies and technologies have to be used simultaneously for your project.

1) Utilize typical energy use patterns for various building types
2) Use statistical databases: performance-based vs. prescriptive approaches
3) Use code-based energy models

Typical Energy Use Pattern Matrix

Building Type	Median Electrical Intensity (kWh/sf-yr)
Education	6.6
Office	11.7
Retail (except mall)	8.0
Food Sales	58.9
Food Service	28.7
Lodging	12.6

4) Reduce energy use: reduce the use of artificial HVAC and lighting; use energy-efficient equipment with feedbacks and controls
5) Consider building orientation and improve building envelope performance
6) Use Energy Star appliances and energy-efficient equipment
7) Measure and verify building energy performance: build and operate as designed (commissioning, and continuous and retro-commissioning); monitor performance and improvement over time (monitoring and verification)
8) Generate on-site renewable energy: geothermal, solar PV and wind energy
9) Take advantage of the site: passive solar energy, natural ventilation, and passive cooling
10) Buy off-site renewable energy

Specific Technical Information:

EAp1: Fundamental Commissioning of the Building Energy Systems (Required for NC, CS, and Schools)

Purpose:

To confirm that the building's energy systems are built and adjusted correctly, and that they will perform per the basis of design, construction documents, and owner's project requirements.

The advantages of commissioning are confirmation of a system's performance per the owner's requirements, fewer contractor callbacks, lower operating costs, less energy use, better building documentation, and increased occupant productivity.

Credit Path (0 point, mandatory requirements/prerequisites):

Your commissioning team shall complete the following commissioning plan/process per the related USGBC reference guide:

   a. Assign a person to be the <u>Commission Authority (CxA)</u> to review, lead and supervise the completion of the commissioning process.

      1) The CxA shall report the recommendation, findings, and results <u>directly to the owner</u>.

      2) The CxA shall have <u>commissioning authority experience</u> on at least <u>2</u> building projects.

      3) The CxA shall be <u>independent</u> of the project's design and construction management teams, but he or she can be an employee of the firm providing the services. The CxA could be a qualified consultant or employee of the owner.

      4) The CxA can be a qualified member with the required experience on the project's design and construction teams for projects with <u>less than 50,000 gross s.f</u>.

   b. Create and implement a <u>commissioning plan</u>.

   c. Create and implement commissioning <u>requirements</u> into the construction documents.

   d. <u>Confirm</u> the performance and installation of the system.

   e. Finish a summary commissioning <u>report</u>. The report shall include:

      1) <u>Executive summary</u>

      2) <u>Test results and evaluation</u>

      3) <u>History of deficiencies</u>

      4) <u>Confirmation from the CxA whether the systems meet the Basis of Design (BOD), Owner's Project Requirements (OPR) and Constructions Documents (CD)</u>

     If you are pursuing credit for EAc3, Enhanced Commissioning, also include:

      5) <u>Submittal process summary</u>

      6) <u>Design review summary</u>

      7) <u>Operations and maintenance documentation, and training process</u>

      8) <u>As-built drawings</u>

   f. The design team needs to create the <u>Basis of Design (BOD)</u>. The owner needs to document the <u>Owner's Project Requirements (OPR)</u>. The CxA needs to review these for clarity and completeness. The design team and owner are responsible for updating their own documents.

You shall finish the commissioning process activities for <u>at least</u> the following systems:

   a. Wind, solar or other <u>r</u>enewable energy systems.

   b. Daylighting and lighting controls (<u>E</u>lectrical).

   c. <u>M</u>echanical and passive HVAC and refrigeration systems and related controls.

   d. Domestic hot water systems (<u>P</u>lumbing).

Mnemonics: <u>REMP</u> (See underlined letters above)

Submittals:

   1) <u>Updated commissioning plan</u> at milestones of the project, including but not limited to the <u>DD</u> (Design Development) phase, the <u>CD</u> (Construction Documents) phase, and <u>right before the kick-off meeting</u> with the GC (general contractor).

   2) <u>A systems list</u> with commissioned systems noted

   3) <u>Confirmation</u> that CxA has <u>documented experience</u> on a minimum of <u>2</u> building projects

   4) Copies of <u>OPR</u>, <u>BOD</u>, commissioning <u>specifications and report</u>, and system <u>manual</u>

Synergies:

Fundamental commissioning may contribute to the following credits:

- SSc8: Light Pollution Reduction

- WEc1: Water-Efficient Landscaping

- WEc2: Innovative Wastewater Technologies
- WEc3: Water Use Reduction

- EAc1: <u>Optimize</u> Energy Performance
- EAc2: On-Site Renewable Energy
- EAc5: Measurement and Verification

- IEQp1: Minimum IAQ Performance
- IEQc1: Outdoor Air Delivery Monitoring
- IEQc2: Increased Ventilation
- IEQc5: Indoor Chemical and Pollutant Source Control Products
- IEQc6: Controllability of Systems
- IEQc7: Thermal Comfort

EAp1 defines a <u>minimum</u> threshold for commissioning. The GBCI awards <u>additional</u> verification and rigor in the following related credit:

- EAc3: Enhanced Commissioning

Possible Strategies and Technologies:
Encourage owners to seek qualified individuals with high levels of experience in the following areas to lead the commissioning process:
a. Process management and commissioning planning.
b. The design, installation, and operation of the energy system.
c. Knowledge of energy systems' automation control.
d. Practical experience with start-up, performance, interaction, troubleshooting, balancing, testing, maintenance, and operating procedures of the energy systems.

Encourage owners to include building envelope systems, water-using systems, etc., in the commissioning plan if applicable. A proper building envelope can reduce energy consumption and improve indoor air quality and occupant comfort. Owners can save a lot of money and improve indoor air quality if they include <u>building envelope</u> commissioning, even though it is not required to be commissioned by LEED.

See the USGBC reference guide for guidance on:
a. Commissioning <u>reports</u>
b. Commissioning <u>plans</u>
c. Commissioning <u>specifications</u>
d. Basis of Design (<u>BOD</u>)
e. Owner's Project Requirements (<u>OPR</u>)
f. Performance <u>verification</u> documentation

Extra Credit (Exemplary Performance): None

Project Phase: Construction Administration

<u>**LEED Submittal Phase:**</u> <u>Construction</u>

Related Code or Standard: None

Responsible Party: Commissioning Authority/Agent, Owner, and Contractor

EAp2: Minimum Energy Performance (Required for NC, CS, and Schools)

Purpose:
1) To set up the minimum energy efficiency level for the proposed building and systems.
2) To avoid excessive energy use and related economic and environmental impact.

Credit Paths (0 points, mandatory requirements/prerequisites):

1) Whole Building Energy Simulation for NC, CS, and Schools

Compared with the baseline building rating per ANSI/ASHRAE/IESNA Standard 90.1-2007 (with errata but without addenda), you need to show a 10% improvement for a new building, or a 5% improvement for a major renovation to an existing building in your proposed building performance rating.

This comparison needs to be a whole building project simulation per the Building Performance Rating Method in Appendix G of the standard.

Note: A project team can choose to use the ASHRAE approved addenda for this credit ONLY if the team applies the addenda consistently across ALL LEED credits.

Your design must:
a. Meet the mandatory provisions (sections 5.4, 6.4, 7.4, 8.4, 9.4 and 10.4) in ASHRAE/IESNA Standard 90.1-2007 (with errata but without addenda);
b. Include ALL of the energy cost associated with and within the building project;
c. Be compared with a baseline building that meets Appendix G of ASHRAE/IESNA Standard 90.1-2007 (with errata but without addenda). Your default process energy cost is 25% of the baseline building's overall energy cost. If the cost is less than 25%, your LEED submittal must justify that your process energy inputs are correct.

AND
d. You need to do 4 baseline design simulations (1 for each orientation), and 1 proposed design simulation.
e. You need to provide sensors for break rooms, classrooms, and conference rooms.
f. You can improve your building's energy performance by these methods: reduce the building footprint; reduce the energy demand; install a controlling sensor for the HVAC systems; use harvested energy (natural ventilation, building orientation, window locations); increase the HVAC systems' efficiency; recover waste energy (gray water heat load, exhaust air waste energy, cogeneration).
g. This path does not apply to warehouses, manufactured homes, and buildings with three stories or less. It covers mechanical systems, service water heating (plumbing), lighting and power systems (electrical), and the building envelope.

In this analysis, you should include the following as part of **process energy**:
Refrigeration and kitchen cooking, laundry (washing and drying), elevators and escalators, computers, office and general miscellaneous equipment, lighting not included in the lighting power allowance (such as lighting as part of the medical equipment), and other uses like water pumps, etc.

You should include the following as **regulated (non-process) energy**:
HVAC, exhaust fans and hood, lighting for interiors, surface parking, garage parking, building façade and grounds, space heating and service water heating, etc.

To meet EAp1, you should use the same process loads for your proposed building performance rating and the baseline performance building rating. You can use the Exceptional Calculation Method (ANSI/ASHRAE/IESNA 90.1-2007 G2.5) to record measures of reducing process loads. You should include assumptions made for both your proposed design and the base building, and related supporting empirical and theoretical information in your documentation of process load energy savings.

If your project is in California, you can use Title 24-2005, Part 6 instead of ANSI/ASHRAE/IESNA 90.1-2007 for this path.

OR

2) **Prescriptive Compliance Path for NC and CS (ASHRAE Advanced Energy Design Guide)**

You must follow all applicable criteria for your building's climate zone per the ASHRAE Advanced Energy Design Guide.
You need to meet the prescriptive measures of ASHRAE Advanced Energy Design Guide suitable for your project, as listed below, and comply with the following:
a. ASHRAE Advanced Energy Design Guide for Small Office Buildings 2004
   - Your building should be **office** occupancy.
   - Your building should be less than 20,000 s.f.
b. ASHRAE Advanced Energy Design Guide for Small Retail Buildings 2006
   - Your building should be **retail** occupancy.
   - Your building should be less than 20,000 s.f.
c. ASHRAE Advanced Energy Design Guide for Small Warehouses and Self Storage Buildings 2008
   - Your building should be **Warehouses or Self Storage** occupancy.
   - Your building should be less than 50,000 s.f.

OR

2) **Prescriptive Compliance Path for Schools (ASHRAE Advanced Energy Design Guide)**
You must follow all applicable criteria for your building's climate zone per the Advanced Energy Design Guide.

You need to meet the prescriptive measures of the ASHRAE Advanced Energy Design Guide for K-12 School Buildings. Your project shall be less than 200,000 s.f.

OR

3) **Prescriptive Compliance Path for <u>NC and CS and Schools</u>: Advanced Buildings™ Core Performance™ Guide**

You need to meet the <u>Advanced Buildings™ Core Performance™ Guide</u>, published by the <u>New Buildings Institute</u>.

   a. This guide is for buildings <u><100,000 s.f.</u>

   b. It is not applicable to <u>warehouse, healthcare, or laboratory</u> projects.

   c. Your project must comply with Section <u>1</u> of <u>Design Process Strategies</u> and Section <u>2</u> of the requirements of the <u>Advanced Buildings™ Core Performance™ Guide</u>.

   d. All <u>school, office, public assembly and retail</u> projects less than 100,000 s.f. must comply with Section <u>1</u> and Section <u>2</u> mentioned above.

   e. Other projects less than 100,000 s.f. shall meet the basic requirements of the <u>Advanced Buildings™ Core Performance™ Guide</u>.

   f. Laboratory, warehouse and health care projects are <u>not</u> qualified for this path.

Submittals:

   1) Confirm the project meets <u>ASHRAE</u> (including addenda) requirements, and keep copies of related <u>forms</u>

   2) Determine the project <u>climate zone</u>

   3) Calculate each type of <u>energy use</u>

   4) Keep <u>a list of energy end uses</u> for both the design case and baseline case

   5) When using a **computer simulation**, follow <u>Appendix G of ASHRAE 90.1-2007</u> (or equivalent local code), and keep copies of the final report. The final report shall show the <u>annual energy cost</u> for both the design case and the baseline case.

   6) When using the **prescriptive compliance** path, gather <u>paperwork</u> showing that the project meets all requirements.

Synergies:

Most LEED systems cover building energy efficiency in two places:

- EAp2: <u>Minimum</u> Energy Performance
- EAc1: <u>Optimize</u> Energy Performance

You can reduce energy consumption for a building by exceeding the building code requirements for HVAC systems, lighting and the building envelope. You can also optimize the exterior lighting or use roofing materials suitable to the local climate to reduce energy use. See credits below:

- EAc1: <u>Optimize</u> Energy Performance
- SSc7.2: Heat Island Effect: Roof
- SSc8: Light Pollution Reduction

Renewable energy may contribute to energy savings:

- EAc2: On-Site Renewable Energy
- EAc6: Green Power

The project team must carefully coordinate IEQ and building energy performance. Do not compromise the well-being or health of the building occupants to save energy. You may need to use extra energy to increase ventilation, and extra energy may cause water and air pollution. You may use economizer and/or heat-recovery ventilation to reduce energy use. Review and coordinate these credits carefully:

- IEQp1: Minimum IAQ Performance
- IEQc1: Outdoor Air Delivery Monitoring
- IEQc2: Increased Ventilation
- IEQc6: Controllability of Systems
- IEQc7: Thermal Comfort
- IEQc8: Daylight and Views

Reducing water use, especially hot water use, may save energy. See the following credits for strategies:
- WEc3: Water Use Reduction:
- WEc4, Process Water Use Reduction

Possible Strategies and Technologies:
1) Assess your building's energy performance and find out the most cost-effective energy efficiency measures with the help of <u>a computer simulation model to quantify energy performance</u> by comparing it to a baseline building. You can design your building systems and envelope to achieve the best energy performance.
2) Correlate local code performance with <u>ASHRAE 90.1-2007 (with errata but without addenda)</u> if it has shown textural and quantitative equivalence per the U.S. Department of Energy standard process for commercial energy code determination. See link below for more information on the <u>DOE</u> process for commercial energy code determination:
http://www.energycodes.gov/

Extra Credit (Exemplary Performance): None

Project Phase: Schematic Design

<u>LEED Submittal</u> Phase: <u>Design</u>

Related Code or Standard:
1) <u>ANSI/ASHRAE/IESNA Standard 90.1-2007</u>: Energy Standard for Buildings Except Low-Rise Residential
2) <u>ASHRAE Advanced Energy Design Guide for Small Office Buildings 2004</u>
3) <u>ASHRAE Advanced Energy Design Guide for Small Warehouses and Self Storage Buildings 2008</u>
4) <u>ASHRAE Advanced Energy Design Guide for K-12 School Buildings</u>
5) <u>New Buildings Institute, Advanced Buildings™ Core Performance™ Guide</u>
6) <u>Energy Star Program, Target Finder Rating Tool</u>

Responsible Party: <u>MEP Engineer</u> and Owner

EAp3: Fundamental Refrigerant Management (Required for NC, CS, and Schools)

Purpose:
To minimize the depletion of stratospheric ozone.

Credit Paths for NC, CS, and Schools (0 point, mandatory requirements/ prerequisites):
1) The HVAC&R (R is for Refrigeration) system of your <u>new</u> base building should <u>NOT</u> use CFC-based refrigerants.

   OR
2) Finish a complete <u>CFC phase-out conversion before project completion</u> if you are reusing the HVAC equipment of an <u>existing</u> base building. The GBCI will consider the phase-out plans continuing beyond the job completion date on their merits.

Note:
1) Do NOT include HVAC units with less than 0.5 lb of refrigerant and small water coolers and standard refrigerators in the base building calculation.
2) Select the HVAC units with small ODP (Ozone Depletion Potential), GWP (Global Warming Potential), and short environmental life.
3) HVAC units with payback times longer than 10-years are non-feasible units.
4) The manufacture of HVAC units containing CFCs was stopped in 1995 in the US and these units will be phased out from existing buildings in the U.S. by 2011.

Submittals:
1) For major renovations, create and track the <u>phase-out plan</u>
2) Gather manufacturer's data showing the <u>type of refrigerant</u> used by the base building HVAC&R system

Synergies:
EAp3 has the minimum requirements for refrigerant selection. See "EAc4: Enhanced Refrigerant Management" for further requirements.

Possible Strategies and Technologies:
1) You should use <u>no CFC-based refrigerants</u> for <u>new</u> buildings.
2) For HVAC systems in <u>existing</u> buildings, you should create an <u>inventory</u> listing the equipment using CFC refrigerants and supply a schedule to <u>replace</u> these refrigerants.

Extra Credit (Exemplary Performance): None

Project Phase: Design Development

<u>LEED Submittal</u> Phase: <u>Design</u>

Related Code or Standard:
1) <u>U.S. EPA Clean Air Act, Title VI, Section 608, Compliance with the Section 608 Refrigerant Recycling Rule</u>
2) <u>Montreal Protocol 1987</u>

Responsible Party: <u>MEP Engineer</u> and Owner

EAc1: Optimize Energy Performance (<u>1-19</u> points for NC and Schools, <u>3-21</u> points for CS)

Purpose:
1) To avoid excessive energy use and reduce related economic and environmental impacts, and
2) To achieve better energy performance above the baseline in the prerequisite standard.

Credit Paths for NC, CS, and Schools:
You can choose one of the three paths below. If you use any one of the following choices to document achievement, you are assumed to comply with EAp2.

1) **Whole Building Energy <u>Simulation</u> (<u>1-19</u> points for NC and Schools, <u>3-21</u> points for CS).**
Compared with the baseline building rating per <u>ANSI/ASHRAE/IESNA Standard 90.1-2007</u>, you need to show a percentage improvement in your proposed building performance rating. This comparison needs to be a whole building project <u>simulation</u> per the Building Performance Rating Method in <u>Appendix G</u> of the standard.

The following are minimum energy savings percentages for getting the points:

Renovating Existing Buildings	New Buildings	Points (CS)	Points (NC and Schools)
8%	12%	3	1
10%	14%	4	2
12%	16%	5	3
14%	18%	6	4
16%	20%	7	5
18%	22%	8	6
20%	24%	9	7
22%	26%	10	8
24%	28%	11	9
26%	30%	12	10
28%	32%	13	11
30%	34%	14	12
32%	36%	15	13
34%	38%	16	14
36%	40%	17	15
38%	42%	18	16
40%	44%	19	17
42%	46%	20	18
44%	48%	21	19

Here is the **tip on how to understand and memorize the above table:**
As you may notice, for existing building renovations (NC and Schools), you can get 1 point for the first 8% of energy savings, and then 1 additional point for every 2% increase in energy savings, up to 19 points maximum. For new buildings (NC and Schools), you can get 1 point for the first 12% of energy savings, and then 1 additional point for every 2% increase in energy savings, up to 19 points maximum. You can look for similar rules to memorize points for CS projects.

Per the Appendix G of ASHRAE/IESNA Standard 90.1-2007 (with errata but without addenda), you must include ALL of the energy costs associated with the building project when you do the energy analysis via the Building Performance Rating Method. To get points through this credit, your design must:
a. Meet the mandatory provisions (sections 5.4, 6.4, 7.4, 8.4, 9.4 and 10.4) in ASHRAE/IESNA Standard 90.1-2007 (with errata but without addenda);
b. Include ALL of the energy costs associated with and within the building project;
c. Be compared with a baseline building that meets Appendix G of ASHRAE/IESNA Standard 90.1-2007 (with errata but without addenda). Your default process energy cost is 25% of the baseline building's overall energy cost. If your cost is less than 25%, your LEED submittal must justify that your process energy inputs are proper.
d. Include 4 baseline design simulations (1 for each orientation), and 1 proposed design simulation.
e. Provide sensors for break rooms, classrooms, and conference rooms.
f. Improve your building's energy performance by these methods: reduce the building footprint; reduce the energy demand; install controlling sensors for HVAC systems; use harvested energy (natural ventilation, building orientation, window locations); increase HVAC system efficiency; recover waste energy (gray water heat load, exhaust air waste energy, cogeneration).

This credit path does not apply to warehouses, manufactured homes, and buildings with three stories or less. It covers mechanical system, service water heating (plumbing), lighting system and power system (electrical), and the building envelope.

In this analysis, include the following as part of **process energy**:
Refrigeration and kitchen cooking, laundry washing and drying, elevators and escalators, computers, office and general miscellaneous equipment, lighting not included in the lighting power allowance (such as lighting as part of the medical equipment), and other uses like water pumps, etc.

You should include the following as **regulated (non-process) energy**:
HVAC, exhaust fans and hoods, lighting for interiors, surface parking, garage parking, building façade and grounds, space heating, and service water heating, etc.

To get EAc1, you should use the same process loads for your proposed building performance rating and the baseline performance building rating. You can use the Exceptional Calculation Method (ANSI/ASHRAE/IESNA 90.1-2007 G2.5) to record measures of reducing process loads. Include the assumptions made for both your proposed design and the base building, as well as related supporting empirical and theoretical information in your documentation of process load energy savings.

OR

2) **<u>Prescriptive</u> Compliance Path for NC and CS: ASHRAE Advanced Energy Design Guide (<u>1</u> point)**

Follow all applicable criteria for your building's <u>climate zone</u> per <u>ASHRAE Advanced Energy Design Guide</u>, including those that may be in the tenant's scope of work for CS projects. You may need a tenant lease or a sales agreement to satisfy some of the criteria for CS projects. You need to meet the <u>prescriptive</u> measures of <u>ASHRAE Advanced Energy Design Guide suitable for your project</u> as listed below:

a. ASHRAE Advanced Energy Design Guide for Small <u>Office</u> Buildings 2004
- Your building should be **office** occupancy.
- Your building should be <u>less than 20,000 s.f.</u>

OR

b. ASHRAE Advanced Energy Design Guide for Small <u>Retail</u> Buildings 2006
- Your building should be **retail** occupancy.
- Your building should be <u>less than 20,000 s.f.</u>

OR

c. ASHRAE Advanced Energy Design Guide for Small Warehouse and Self-Storage Building 2008
- Your building should be **<u>Small Warehouse or Self-Storage</u>** occupancy.
- Your building should be <u>less than **50,000** s.f.</u>

OR

2) **<u>Prescriptive</u> Compliance Path for <u>Schools</u>: ASHRAE Advanced Energy Design Guide for K-12 School Building (<u>1</u> point)**
- Your project (or building<u>s</u>) should be <u>less than **200,000** s.f.</u>

OR

3) **<u>Prescriptive</u> Compliance Path for NC and CS and Schools: Advanced Buildings™ Core Performance™ Guide (<u>1-3</u> points)**
You need to meet the <u>Advanced Buildings™ Core Performance™ Guide</u>, published by the New Buildings Institute.
- This guide is for buildings <u><100,000 s.f.</u>
- This guide is not applicable to <u>warehouse, healthcare, or laboratory</u> projects.
- Your projects must comply with Section <u>1</u>, Design Process Strategies and Section <u>2</u>, Core Performance™ Requirements.
- You can get **<u>1</u> point** for all <u>school, office, public assembly and retail</u> projects that are less than <u>100,000 s.f.</u> and comply with Section 1 and 2 of the guide.
- You can get **2 additional points** for implementing strategies listed in Section 3 of the guide: <u>1 extra point for every 3 strategies</u> implemented, except the following strategies: <u>3.1, Cool Roofs; 3.8, Night Venting; 3.13, Additional Commissioning</u>, because these strategies are already covered by other aspects of the LEED program.

Submittals:
Same as EAp2

Synergies:
Same as EAp2, except EAc1 is replaced by EAp2.

Possible Strategies and Technologies:
1) With the help of <u>a computer simulation model</u>, you can <u>quantify energy performance</u> by comparing it to the baseline building, assess your building's energy performance, find out the most cost-effective energy efficiency measures, and design your building systems and envelope to achieve the best energy performance.
2) You can correlate local code performance with <u>ASHRAE 90.1-2007 (with errata but without addenda)</u> if it has shown textural and quantitative equivalence per at least the U.S. Department of Energy standard process for commercial energy code determination. See link below for more information on the <u>DOE</u> process for commercial energy code determination:
http://www.energycodes.gov/
3) Use **Combined Heat and Power** systems (**CHP**). They <u>capture the heat</u> that would have been wasted in the process of generating electricity via fossil fuel. They are much <u>more efficient</u> than separate thermal systems and central power plants, <u>reduce peak demand</u>, generate <u>fewer emissions</u>, <u>reduce loss</u> in electricity transmission and distribution, and <u>release</u> electrical grid <u>capacity</u> for other uses.

Extra Credit (Exemplary Performance):
For Credit **Path #1**: You can get <u>one</u> innovation point if you can save <u>50%</u> of energy for <u>new</u> buildings OR save <u>46%</u> of energy for <u>existing</u> buildings.

For Credit **Path #2 and #3**: None.

For **CS** projects, you can claim additional points via **Credit for Tenant-Implemented Efficiency Measures**. The measures must be included in a tenant's **enforceable** lease agreement. You must provide a copy of the <u>lease agreement</u>, the <u>level of performance</u> to be met by the tenant and a <u>list</u> of such measures.

Project Phase: Schematic Design

LEED Submittal Phase: <u>Design</u>

Related Code or Standard:
1) <u>ANSI/ASHRAE/IESNA Standard 90.1-2007</u>: Energy Standard for Buildings Except Low-Rise Residential, and Informative Appendix G, Performance Rating Method
2) <u>ASHRAE Advanced Energy Design Guide for Small Office Buildings 2004</u>
3) <u>ASHRAE Advanced Energy Design Guide for Retail Buildings 2006</u>
4) <u>ASHRAE Advanced Energy Design Guide for Small Warehouses and Self Storage Buildings 2008</u>
5) <u>ASHRAE Advanced Energy Design Guide for K-12 School Buildings</u>
6) <u>New Buildings Institute, Advanced Buildings™ Core Performance™ Guide</u>

Responsible Party: <u>MEP Engineer</u> and Owner

EAc2: On-Site Renewable Energy (1-7 points for NC and Schools, 4 points for CS)

Purpose:
To reduce energy created with fossil fuels and minimize the related economic and environmental impacts by recognizing and encouraging increasing levels of self-supplied on-site renewable energy.

Credit Path:
To get this credit, reduce the building energy cost with the help of on-site renewable energy systems. Divide the energy produced via on-site renewable energy systems by your building's annual energy cost. Then use this percentage as well as the table below to determine the number of points that you will get.

You can use the Department of Energy (DOE) Commercial Building Energy Consumption Survey (CBECS) database or your building's annual energy cost calculated in EAc1 to estimate your electricity use. You can find tables for different building types in the reference guide.

For NC and Schools

% On-site Renewable Energy	Points
1%	1
3%	2
5%	3
7%	4
9%	5
11%	6
13%	7

Note: This credit also creates an opportunity for schools to include sustainable technologies as part of their curriculums.

For CS

% On-site Renewable Energy	Points
1%	4

Submittals:
1) Documentation for total annual energy generation, on-site renewable energy type, and backup energy source
2) Calculate the amount of energy from each on-site renewable source
3) Keep documents for any incentives to support the on-site renewable energy systems installation.

Synergies:
On-site renewable energy equipment typically has a small impact on other credits. It requires measurement and verification, commissioning, and it also affects building energy performance. It can also reduce the amount of green power purchased. EAc2 may contribute to the following credits:
- EAp1: Fundamental Commissioning of the Building Energy Systems
- EAp2: Minimum Energy Performance

- EAc1: <u>Optimize</u> Energy Performance
- EAc5: Measurement and Verification
- EAc6: Green Power

Possible Strategies and Technologies:

1) Evaluate your project for <u>biogas, biomass,</u> or any other <u>renewable and non-polluting energy</u> strategies.

2) Use <u>net metering</u> to send extra on-site renewable energy back to the grid. Coordinate with your local utility company regarding potential rebate and incentive programs.

3) Use <u>virtual energy (annual energy cost divided by annual energy consumption)</u> and <u>local utilities rates</u> to assign a dollar value.

4) <u>Geothermal heating</u> systems, <u>wind energy</u> systems, <u>wave and tidal</u> systems, <u>biofuel-base electrical</u> systems, <u>low-impact hydro electric</u> systems, <u>photovoltaic</u> systems, and <u>solar thermal</u> systems are **eligible systems** for this credit; <u>architectural features, daylighting and passive solar</u> strategies, <u>geo-exchange</u> systems <u>(ground source heat pumps)</u> and <u>purchased/green power (it should be part of EAc6)</u> are **ineligible on-site systems** for this credit. <u>Architectural features, daylighting and passive solar</u> strategies, and <u>geo-exchange</u> are <u>non-eligible</u> because they are already covered in EAp2 and may be considered under EAc1.

5) You <u>can</u> include the following biofuels as part of **renewable energy** for this credit: <u>agricultural waste or crops, organic waste including animal waste, landfill gas, and mill residues and other untreated wood waste.</u>

6) You <u>cannot</u> include the following biofuels as part of **renewable energy** for this credit: <u>municipal solid waste combustion, wood coated with formica, plastic or paint, preserved wood treated with materials containing arsenic, chlorine compounds, chromate copper arsenate, halide compounds, or halogens.</u> If <u>over 1%</u> of the preserved wood fuel has been treated with these compounds, the energy system becomes ineligible.

Extra Credit (Exemplary Performance):

For NC and Schools

You can get one innovation point by achieving <u>15%</u> of On-site Renewable Energy.

For CS

You can get one innovation point by achieving <u>5%</u> of energy use from on-site renewable energy sources.

Project Phase: Schematic Design

LEED Submittal Phase: <u>Design</u>

Related Code or Standard:

1) <u>ANSI/ASHRAE/IESNA Standard 90.1-2007, Energy Standard for Buildings Except Low-Rise Residential (with errata but without addenda)</u>

2) <u>Department of Energy (DOE)</u>

3) <u>Commercial Building Energy Consumption Survey (CBECS)</u>

4) <u>EIA (Energy Information Administration)</u>

Responsible Party: <u>MEP Engineer</u>

EAc3: Enhanced Commissioning (2 points for NC, Schools, and CS)

Purpose:
To designate an <u>independent</u> Commissioning Authority (CxA) before the start of the construction documents (CD) phase, and verify the building systems' performance, and then perform additional activities.

Credit Path for NC, Schools, and CS:
To get this credit, you need to:
   a. Perform ALL requirements of EAp1, AND
   b. Perform (or have a contract to perform) the following commissioning process activities that are <u>above and beyond</u> the requirements of EAp1 and comply with the latest version of the related USGBC reference guide:
      1) Review contractor <u>s</u>ubmittals related to systems commissioned to make sure they meet the **Basis of Design (BOD)** and **Owner's Project Requirements (OPR)**. This needs to be concurrent with Architect/Engineer reviews. The submittals shall also be sent to the owner and the design team.
      2) Make a system <u>m</u>anual to provide information for future system operators to be able to understand and optimally operate the commissioned systems.
      3) <u>A</u>rrange to review building operations with occupants and O&M staff within 10 months of the building's substantial completion. Make a plan to resolve outstanding commissioning-related issues.
      4) At least **one** design review of **BOD** and **OP<u>R</u>** and design documents before the mid-construction documents phase, and back check subsequent design submissions against the review comments.
      5) Confirm the completion of the requirements for <u>t</u>raining building occupants and operating personnel.

Note: The CxA shall be <u>independent</u> of the project's design and construction management team and he or she <u>cannot</u> be an employee of the design firm providing service to the project, OR an employee of the construction manager or contractor holding the construction contract. (This is <u>different</u> than EAp1). He or she can be contracted through the <u>design</u> firm. The CxA could be a qualified consultant or employee of the owner.

Mnemonics: <u>SMART</u> (See underlined letters in items #1 through #5 above). The CxA shall perform the #1, #3 and #4 (or <u>S, A</u> and <u>R</u>) of above tasks; the other team member can perform #2 and #5 (or <u>M</u> and <u>T</u>) of the above tasks.

Submittals:
All submittals required by EAp1 PLUS the following:
   1) A written <u>schedule</u> of building operator training
   2) A copy of the CxA's design <u>review</u>, any related designer <u>response</u>, and <u>back-check</u> confirmation

Synergies:
Same as EAp1, except that EAc3 goes above and beyond the minimum threshold for commissioning defined by EAp1.

Possible Strategies and Technologies:
The CxA can be contracted by:

1) The owner (the preferred way)

OR

2) Through construction management firms NOT holding the construction contracts or through the design firm.

You can find detailed guidance on the rigor expected for the following items in the related reference guide:
   a. Systems manual;
   b. Commissioning design review;
   c. Commissioning submittal review.

Extra Credit (Exemplary Performance):
NC, Schools, and CS projects may qualify for <u>one</u> ID bonus point for comprehensive building envelope commissioning. You need to show the protocols and standards of building envelope commissioning.

CS projects may qualify for <u>one</u> ID bonus point for requiring <u>ALL</u> tenant spaces to conduct both the <u>fundamental and enhanced</u> commissioning.

Project Phase: Schematic Design

LEED Submittal Phase: <u>Construction</u>

Related Code or Standard: None

Responsible Party: <u>Commissioning Authority</u>, Owner and Contractor

EAc4: Enhanced Refrigerant Management (<u>2</u> points for NC and CS, <u>1</u> point for Schools)

Purpose:
To minimize direct contribution to global warming through supporting early compliance with <u>Montreal Protocol</u> and reducing ozone depletion.

Credit Paths for NC, Schools, and CS:
   1) <u>Not using refrigerants</u>

OR

   2) Choose HVAC and R and refrigerants that <u>reduce or eliminate</u> the emissions of <u>compound contributing to global warming and ozone depletion</u>. Your base building's equipment needs to <u>comply with the following formula</u>. This is the <u>maximum</u> threshold for the contributions to global warming and ozone depletion:

LCGWP + LCODP × 10^5 ≤ 100

Note:

$$LCGWP = \frac{GW\Pr \times (Lr \times Life \times Mr) \times Rc}{Life} \quad LCODP = \frac{OD\Pr \times (Lr \times Life \times Mr) \times Rc}{Life}$$

LCGWP: Lifecycle Global Warming Potential (lb CO_2/Ton-Year)
LCODP: Lifecycle Ozone Depletion Potential (lb CFC11/Ton-Year)
GWPr: Global Warming Potential of Refrigerant (0 to 12,000 lb CO_2/lbr)
ODPr: Ozone Depletion Potential of Refrigerant (0 to 0.2 lb CFC 11/lbr)
Lr: Refrigerant Leakage Rate (0.5% to 2.0% default of 2% unless otherwise demonstrated).
Mr: End-of-life Refrigerant Loss (2% to 10%; default of 10% unless otherwise demonstrated).
Rc: Refrigerant Charge (0.5 to 5.0 lbs. of refrigerant per ton of gross ARI rated cooling capacity).
Life: Equipment Life (10 years; default based on equipment type, unless otherwise demonstrated).

For more than one type of equipment, you should use a <u>weighted average</u> of all base building HVAC and R equipment per formula below:

[∑ (LCGWP + LCODP x 10^5) x Qunit]/Qtotal≤100

Note:
Quint = Gross ARI rated cooling capacity of an individual refrigeration or HVAC unit (tons).
Qtotal = Total gross ARI rated cooling capacity of all refrigeration or HVAC.

All paths:
Small water coolers, small HVAC units, standard refrigerators and other cooling equipment containing less than 0.5 lbs. of refrigerant are <u>not</u> included as part of the base building system and are <u>exempt</u> from this credit's requirements.

AND
Do <u>not use fire suppression systems containing ozone-depleting substances (Halons, CFCs, and HCFCs)</u>.

Note:
For global warming potential (<u>GWP</u>) and ozone depletion potential (<u>ODP</u>) and <u>efficiency</u>: <u>CFC>HCFC>HFC</u>

Submittals:
1) A <u>list</u> of building systems containing refrigerants, and refrigerant types, ODP and GWP
2) <u>Manufacturer's data</u>, including the quantity and types of refrigerants used.
3) Documentation from manufacturers or engineers indicating that <u>CFCs, HCFCs, and halons</u> are <u>not</u> used in fire-suppression systems.

Synergies:
EAc4 may contribute to the following credits:
- EAp2: <u>Minimum</u> Energy Performance
- EAp3: Fundamental Refrigerant Management

- EAc1: <u>Optimize</u> Energy Performance
- IEQc7.1: Thermal Comfort-Design
- IEQc7.2: Thermal Comfort-Verification

Possible Strategies and Technologies:
1) You can choose not to use mechanical refrigeration or cooling equipment.
2) When you do use them, you can use the base building refrigeration and HVAC systems for the cycle that <u>minimize direct impact on global warming and ozone depletion</u>.
3) Choose HVAC&R equipment <u>with increased equipment life</u> and minimized refrigerant charge.
4) <u>Prevent leakage</u> of equipment refrigerants through proper maintenance.
5) Do NOT use fire suppression systems containing ozone-depleting substances (<u>Halons, CFCs, and HCFCs</u>).

Extra Credit (Exemplary Performance): None

Project Phase: Design Development

<u>**LEED Submittal Phase:**</u> <u>Design</u>

Related Code or Standard: None

Responsible Party: <u>MEP Engineer</u>

EAc5: Measurement and Verification (<u>3</u> points for NC, <u>2</u> points for Schools and <u>Not Applicable</u> for CS)

Purpose:
To have ongoing accountability of building energy consumption over time.

Credit Paths for NC and Schools:
You need to <u>create and implement a Measurement and Verification (M&V) Plan</u> consistent with
1) **Option B:** Energy Conservation Measure Isolation (for <u>smaller</u> projects) per the *International Performance Measure and Verification Protocol (IPMVP) Volume III: Concepts and Options for Determining Energy Savings in New Construction*, April 2003.

OR
2) **Option D:** Calibrated Simulation (Savings Estimate Method 2, for <u>larger</u> projects) per the *International Performance Measure and Verification Protocol (IPMVP) Volume III: Concepts and Options for Determining Energy Savings in New Construction*, April 2003.

For both paths, the M&V period cover a minimum of **1 year** of post-construction occupancy. Offer a process for corrective action if the M&V plan results show the energy savings have not been achieved.

Submittals:
   For EAc5 (NC & Schools) & EAc5.1 CS
   1) An IPMVP-compliant <u>M&V plan</u>
   2) Show and update the <u>locations of any meters</u> for measurement on a <u>diagram</u>

   For EAc5.2 CS
   1) <u>Guideline</u> on tenant's energy use, including information on how to determine energy use and the related costs

Synergies:
Measurement and verification may help to achieve optimal energy performance and ensure accountability. If you need to obtain an energy performance contract or other funding, you probably should use the International Performance Measurement and Verification Protocol (IPMVP). M&V plans typically include tracking the performance of renewable energy generation systems to identify operational issues.

EAc5 may contribute to the following credits:
   - EAp2: <u>Minimum</u> Energy Performance
   - EAc1: <u>Optimize</u> Energy Performance
   - EAc2: On-Site Renewable Energy

Commissioning and M&V often use the same devices. See the following credits for related criteria:
   - EAp1: <u>Fundamental</u> Commissioning of the Building Energy Systems
   - EAc3: <u>Enhanced</u> Commissioning

Possible Strategies and Technologies:
   1) Create an <u>M&V Plan</u> to evaluate energy systems and/or building performance.
   2) Use engineering <u>analysis</u> and energy <u>simulation</u> to characterize the energy systems and/or building.
   3) Use proper <u>metering</u> to measure the use of energy.
   4) Verify performance by comparing <u>actual</u> performance to <u>predicted</u> performance **(Performance Factor)**; then group them by system or component if needed.
   5) Compare <u>actual</u> performance to <u>baseline</u> performance to track energy **efficiency**.

IPMVP can be used to verify savings related to **energy conservation measures (ECMs)** and strategies. This LEED credit goes above and beyond normal IPMVP M&V goals. M&V can go above and beyond energy systems and ECMs, and energy conservation strategies. IPMVP gives guidance on M&V strategies and their uses. Use these together with trend logging and monitoring of important energy systems, and seek ongoing accountability for building energy performance.

Extra Credit (Exemplary Performance): None

Project Phase: Design Development

<u>**LEED Submittal**</u> **Phase:** <u>Construction</u>

Related Code or Standard:
   1) <u>International Performance Measure and Verification Protocol (IPMVP)</u>, Volume III, Concepts

and Options for Determining Energy Savings in New Construction, April, 2003

Responsible Party: <u>MEP Engineer, Building Controls Designer/Manufacturer</u>

EAc5.1: Measurement and Verification-Base Building (<u>3</u> points for CS)

Note: Detailed discussions have been omitted since this credit is for CS only, and is unlikely to be tested on the LEED Green Associate Exam.

EAc5.2: Measurement and Verification-Tenant Submetering (<u>3</u> points for CS)

Note: Detailed discussions have been omitted since this credit is for CS only, and is unlikely to be tested on the LEED Green Associate Exam.

EAc6: Green Power (<u>2</u> points for NC, Schools and CS)

Purpose:
To encourage the use of renewable, grid-source energy technologies based on zero net pollution.

Credit Paths for NC, Schools, and CS:
You should get a minimum of <u>35%</u> of your building's electricity (based on quantity, NOT the cost) from renewable sources through at least <u>two-year</u> contracts of renewable energy. You should use the definition of a renewable source by <u>Center for Resource Solution's (CRS) Green-e</u> product requirements.

Extra requirements for <u>Schools</u>
You can buy green power on a centralized basis, and assign it to a specific project, but you cannot use it again to gain credit for another project.

Extra requirements for <u>CS</u>
CS projects' electricity is defined as the electricity usage of the CS square footage per the Building Owners and Managers Association's (BOMA) Standards. For this credit, the CS square footage shall not be less that <u>15%</u> of the total building gross square footage. If it is less than 15%, use 15% for calculations.

1) Establish your baseline electricity use per the annual consumption of electricity from the results of EAc1.

OR
2) Establish the predicted electricity use per the <u>Department of Energy's (DOE) Commercial Buildings Energy Consumption Survey (CBECS) database</u>.

Submittals:
1) Keep a record of the signed <u>2-year contract</u> to purchase Green-e certified renewable energy or equivalent
2) For a campus project, keep paperwork showing renewable energy was <u>retained on behalf of the project</u> when others are purchasing the renewable energy.

Synergies:
Using renewable energy may reduce energy costs, see:
- EAc1: <u>Optimize</u> Energy Performance

Commission the on-site renewable energy system. Check the roof for structural stability before installing rooftop equipment.
EAc6 may contribute to the following credits:
- SSc7.2: Heat Island Effect: Roof
- EAp1: <u>Fundamental</u> Commissioning of the Building Energy Systems
- EAc3: <u>Enhanced</u> Commissioning

Possible Strategies and Technologies:
1) Evaluate your building's energy needs and find opportunities to sign a green power contract.
2) Green power comes from <u>wind, solar, biomass, geothermal or low-impact hydro</u> sources. See <u>www.green-e.org</u> for more information on the Green-e program.
3) The power you buy for this credit does <u>NOT</u> need to be Green-e certified. You can use green power from other sources as long as they meet the technical requirements of the <u>Green-e program</u>.
4) Use green tags, tradable renewable certificates (TRCs), renewable energy certificates (RECs) and other green power sources to meet EAc6 requirements as long as they meet the technical requirements of the Green-e Program.

Extra Credit (Exemplary Performance):
You can get one innovation point by purchasing <u>70%</u> of the building's electricity from renewable sources.

Project Phase: Occupation/Operation

<u>LEED Submittal</u> Phase: <u>Construction</u>

Related Code or Standard:
1) <u>Department of Energy (DOE)</u>
2) <u>Commercial Buildings Energy Consumption Survey (CBECS)</u>.
3) <u>Center for Resource Solution (CRS), Green-e Product Certification Requirements</u>

Responsible Party: <u>LEED AP</u> and Owner

EA Summary and Mnemonics:

EA Credit Name	Extra Credit	Responsible Party
EAp1: **F**undamental Commissioning of the Building Energy Systems (Required for NC and CS and Schools)	0	Commissioning Authority/Agent, Owner and Contractor
EAp2: **M**inimum Energy Performance (Required for NC and CS and Schools)	0	MEP Engineer and Owner
EAp3: **F**undamental Refrigerant Management (Required for NC and CS and Schools**)**	0	MEP Engineer and Owner
EAc1: **O**ptimize Energy Performance (1-19 points for NC and Schools, 3-21 points for CS)	For Credit Path #1: You can get one innovation point if you can save 50% of energy for new buildings OR save 46% of energy for existing buildings. For Credit Path #2 and #3: None. For CS projects only, you can claim additional points via Credit for Tenant-Implemented Efficiency Measures. The measures must be included in a tenant's enforceable lease agreement. You must provide a copy of the lease agreement, the level of performance to be met by the tenant and a list of such measures.	MEP Engineer and Owner
EAc2: **O**n-Site Renewable Energy (1-7 points for NC and Schools, 4 points for CS)	For NC and Schools You can get one innovation point by achieving 15% of on-site renewable energy. For CS You can get one innovation point by achieving 5% of energy use from on-site renewable energy sources.	MEP Engineer
EAc3: **E**nhanced Commissioning (2 points for NC, Schools and CS)	NC, Schools and CS projects may qualify for one ID bonus point for comprehensive building envelope commissioning. You need to show the protocols and standards of building envelope commissioning. CS projects may qualify for one ID bonus point for requiring ALL tenant spaces to conduct both the fundamental and enhanced commissioning.	Commissioning Authority, Owner, and Contractor

EAc4: **E**nhanced Refrigerant Management (2 points for NC and CS, 1 point for Schools)	0	MEP Engineer
*EAc5: **M**easurement and Verification (3 points for NC, 2 points for Schools and N/A for CS)	0	MEP Engineer, Building Controls Designer/Manufacturer
*EAc5.1: Measurement and Verification-Base Building (3 points for CS, N/A for NC & Schools)	0	Same as above
*EAc5.2: Measurement and Verification-Tenant Submetering (3 points for CS, N/A for NC & Schools)	0	Same as above
EAc6: **G**reen Power (2 points for NC, Schools and CS)	1 ID bonus point by purchasing 70% of the building's electricity from renewable sources.	LEED AP and Owner

Mnemonics:

Find My Fun!

Oh! Oprah Enters Energy Management Group. (See bold and underlined letters in credit names on table above also). Of course, you can create your own **mnemonics** to help you memorize them.

Note: * indicates prerequisite or credit NOT applicable to all LEED rating systems. Detailed discussions have been omitted since the information is unlikely to be tested on the LEED Green Associate Exam.

D. Materials and Resources (MR)

Overall purpose: The MR category has many purposes including:
1) Use fewer materials (**Reduce**)
2) Use materials with less environmental impact
3) Manage and reduce waste (**Recycle**)
4) Storage and Collection of Recyclables
5) Building **Reuse**: Maintain Existing Walls, Floors and Roof
6) Building Reuse: Maintain Interior Nonstructural Elements
7) Construction Waste Management
8) Materials Reuse
9) Recycled Content
10) Regional Materials
11) Rapidly Renewable Materials

12) Certified Wood

Mnemonics:
My effortless work at SBC
MR Regan Ray Carter (See underlined letters above also)

Core concepts:
1) Manage waste
   - Reduce waste
   - Reuse and divert waste
2) Reduce and reuse materials
   - Reduce materials used
   - Reuse of building and materials
   - Choose rapidly renewable materials

Recognition, regulation and incentives:
1) Recognition
   - Cradle to Cradle, Green Seal and other product certifications
2) Regulation (Requirements and Goals)
   - Rare
   - Internal policy for supply chain and materials management in some organizations
3) Financial Incentives
   - Recycling incentive

Overall strategies and technologies:
Note: Not **all** strategies and technologies have to be used simultaneously for your project
1) **Reduce** waste
2) Purchase sustainable materials
3) **Reuse** and divert waste: **recycle** solid waste and demolition waste
4) Reduce life cycle impact
5) Reduce demand for materials: Implement in design and construction; use new technologies
6) Reuse all or a portion of the existing building
7) Reuse materials: Use refurbished, salvaged, and reclaimed materials; purchase refurbished and reclaimed materials
8) Use **rapidly renewable materials**: wool carpeting, cork flooring, sunflower seed board panels, linoleum flooring, bamboo flooring, cotton batt insulation
   Mnemonics:
   WC on slab (See underlined letters above also)
9) Choose materials with a reduced life cycle impact: regional materials; certified wood; materials containing pre- and post-consumer recycled content

Specific Technical Information:

MRp1: Storage and Collection of Recyclables (Required for NC, Schools, and CS)

Purpose:
To reduce waste generated by your building occupants that is going to landfills.

Credit Path (0 point, mandatory requirements/prerequisites):
Provide <u>an easily accessible designated area</u> for <u>separation, collection, storage</u> and <u>recycling</u> of non-hazardous <u>materials</u>, like **p**aper, corrugated **c**ardboard, **m**etals, **g**lass, and **p**lastics for the entire building.

Mnemonics: <u>P</u>eople <u>C</u>an <u>M</u>ake **G**reen <u>P</u>romises (See **bold** and <u>underlined</u> letters at the last part of the sentence listing the recycled materials. I use bold font here also because the underline for letter "g" is hard to read.)

Note: These <u>five</u> materials required for <u>recycling</u> (as listed above) make up <u>59%</u> of the total municipal solid waste stream.

Food scraps (12%) and yard trimmings (13%) make up for <u>25%</u> of the total. The GBCI encourages you to <u>compost</u> these types of waste on-site if possible.

The remaining waste is wood (6%), textiles, leather and rubber (7%) and other (3%).

Submittals:
1) Maintain a recycling plan showing the recycling area's size and accessibility to facility, staff and occupants. Provide adequate size
2) Create floor plans, site plans, etc. to highlight all storage areas

Synergies:
1) The project team can use <u>displays and signage</u> to inform visitors and occupants about on-site recycling and seek an ID credit for educational outreach.
2) For CS projects, use tenant <u>guidelines</u> to address recycling policy and procedure. Encourage the **3 R's**: <u>reduce, reuse, and recycle</u>.

Possible Strategies and Technologies:
1) <u>Coordinate with all collection services</u> to assure effective size and function of the designated recycle areas.
2) Use these methods: <u>separation, collection, storage</u>.
3) You can use <u>aluminum can crushers, card-board balers, collection bins and recycling chutes (collection sources)</u> to make the recycling program more effective.
4) For **CS** projects, you can incorporate the building's recycling policy into the <u>tenant guidelines</u>, and encourage tenants to <u>reduce</u> and <u>reuse</u> before <u>recycling</u>. You also can consider the waste management and maintenance practices for the entire building.

If a <u>landlord</u> is providing cleaning services for all tenants, then he can control the <u>procedures and space</u> needed for storing, removing, and hauling recycled materials. He needs to <u>coordinate the</u>

space needed and the frequency of collecting the recycled materials and waste from the tenant space.

If tenants are handling their own cleaning services, then the landlord needs to provide adequate space for storing recycled materials and clear instructions in the tenant guidelines regarding use of the space. The landlord may provide several collection points at the central building area for tenants' convenience in large buildings or buildings with multiple floors.

5) For institutional projects like **schools**, a recycling team run by the occupants or students to take materials from small collection bins to a central collection location can provide chances for collaboration and learning.

6) **Guideline for Recycling Area**

Commercial Building Area (s.f.)	Min. Recycling Area (s.f.)
0-5,000	82
5,001-15,000	125
15,001-50,000	175
50,001-100,000	225
100,001-200,000	275
200,000 or more	500

Extra Credit (Exemplary Performance): None

Project Phase: Schematic Design

LEED Submittal Phase: Design

Related Code or Standard:
   1) California Integrated Waste Management Board 1999 (CIWMB 1999)

Responsible Party: Architect and Owner

*MRc1.1 for NC and Schools, *MRc1 for CS: Building Reuse: Maintain Existing Walls, Floors and Roof (1-3 points for NC, 1-2 points Schools, and 1-5 points CS)

Purpose:
To fully take advantage of existing buildings, save resources, preserve cultural heritage, minimize waste and new buildings' environmental impact, and minimize materials transportation and manufacturing cost.

Credit Paths for NC, Schools, and CS:
Keep the existing building envelope and structure, including walls, roof and floors (based on surface area). You can include exterior framing and skin, roof decking and structural floor, but you cannot include non-structural roofing materials and window assemblies. You should also exclude hazardous

materials remediated in your job from the percentage calculation of the materials you keep.

For **NC and School** projects, if you are doing an addition and the area of the new addition is more than two times the area of the existing building, your project is not eligible for this credit.

For **CS** projects, if you are doing an addition and the area of the new addition is more than six times the area of the existing building, your project is not eligible for this credit.
Do not include MEP units and elevator machines in the calculations.

NC

% of Building Reuse	Points
55%	1
75%	2
95%	3

Schools

% of Building Reuse	Points
75%	1
95%	2

CS

% of Building Reuse	Points
25%	1
33%	2
42%	3
50%	4
75%	5

Submittals:
1) A list of building shell elements, including IDs for the elements and the total areas of reused, existing, and new elements
2) An explanation as to why any existing building elements were not reused

Synergies:
Assess the site early to determine which materials and areas can be reused. Incorporate a reuse strategy into the initial design charrettes.

A reuse management plan can determine if the reuse materials meet MRc1 requirements. If the materials have not been counted toward MRc1, they can contribute to:
MRc2: Construction Waste Management

MRc1.1 is also related to:
MRc3: Materials Reuse

Possible Strategies and Technologies:
1) Keep existing building when possible, but remove hazardous materials.
2) Upgrade windows, plumbing fixtures, and mechanical systems, etc., to save water and energy.

Extra Credit (Exemplary Performance):
For **CS** projects, you can get one ID extra point for reusing 95% or more of the building.

Project Phase: Pre-Design

LEED Submittal Phase: Construction

Related Code or Standard: None

Responsible Party: Architect, Contractor and Owner

MRc1.2: Building Reuse: Maintain Interior Nonstructural Elements (1 point for NC and Schools. Not Applicable to CS)

Purpose:
To fully take advantage of existing buildings, save resources, preserve cultural heritage, minimize waste and new buildings' environmental impact, and minimize material transportation and manufacturing costs.

Credit Path for NC and Schools:
Keep 50% or more of the existing nonstructural elements like doors, floor coverings, ceiling systems and interior walls (based on surface area).

If you are doing an addition and the area of the new addition is more than two times the area of the existing building, your project is not eligible for this credit.

Do not include MEP units and elevator machines in the calculation.

Submittals:
1) A list of interior nonstructural elements, including IDs for elements and the total areas for reused, existing and new elements

Synergies:
MRc1.2 is related to the following credits:
- MRc1.1: Building Reuse
- MRc2: Construction Waste Management
- MRc3: Materials Reuse

If there are not enough reuse materials to achieve MRc1.1, then you can count the materials toward MRc2, or MRc3 (but not both).

Possible Strategies and Technologies:
1) Keep existing buildings when possible, but remove hazardous materials.

Extra Credit (Exemplary Performance): None

Project Phase: Pre-Design

LEED Submittal Phase: Construction

Related Code or Standard: None

Responsible Party: Architect, Contractor, and Owner

MRc2: Construction Waste Management (1-2 points for NC, Schools, and CS)

Purpose:
1) To divert land-clearing, demolition and construction debris from disposal in incinerators and landfills.
2) To redirect reusable materials to proper sites, and recover recyclable resources to be reused in the manufacturing process.

Credit Path for NC, Schools, and NC:
Salvage and/or recycle non-hazardous demolition and construction debris. You need to create and implement a plan to identify the materials to be reused, recycled and recovered from disposal and whether they will be comingled or sorted on-site. You cannot count land-clearing debris and excavated soils as part of the material for this credit. You can calculate by volume or weight, but you need to be consistent throughout the whole job.

% of Salvaged or Recycled Materials by Weight or Volume	Points
50%	1
75%	2

Submittals:
1) A log of all types of construction waste, the quantities of each type that were landfilled or diverted, and percentage of waste diverted
2) A construction waste management plan, including diversion goals, related construction materials and debris to be diverted, protocols for implementation, and responsible parties

Synergies:
MRc2 is related to the following credits:
- MRc1: Building Reuse
- MRc3: Materials Reuse

If there are not enough reuse materials to achieve MRc1.1, then you can count the material toward MRc2, or MRc3 (but not both).

If your project has asbestos, lead or other contaminated substances, they should be remediated per EPA requirements. See the following credit:
- SSc3: Brownfield Redevelopment

Possible Strategies and Technologies:
1) Set goals for diverting the waste and create and implement a construction management plan to achieve the goals.
2) Many building materials can be salvaged, refurbished, reused or recycled: glass, plastic, wood, metal, brick, cardboard, insulation, carpet, acoustic ceiling tiles, and gypsum wallboard, etc.
3) Salvageable materials include beams, cabinetry, flooring, and paneling.
4) Include on-site salvaged and reused materials if they have NOT been used for MRc3.1 or MRc3.2.
5) Select an area of your job site for comingled or segregated materials for recycling; keep track of your diverting and recycling effort.
6) Select recyclers and haulers to handle recycled materials.
7) Salvage your materials on-site or donate your materials to charity.
8) Exclude soils, vegetation and rocks from the site, but include MEP equipment (Note: MEP equipment can ONLY be included in MRc2).

Extra Credit (Exemplary Performance):
You can get one innovation point for MRc2 by salvaging and/or recycling a minimum of 95% of non-hazardous demolition and construction debris.

Project Phase: Construction Documents

LEED Submittal Phase: Construction

Related Code or Standard: None

Responsible Party: Contractor

MRc3: Materials Reuse (1-2 points for NC and Schools, and 1 point for CS)

Purpose:
To reuse building products and materials to reduce waste and save original/virgin materials, also to reduce the environmental impact related to processing and extracting original materials.

Credit Path for NC and Schools:
A total of 5% or 10% or more (based on cost) of the materials that you use should be reused, refurbished or salvaged materials. The approved materials can be from off-site sources.

You can only include permanently installed materials for your job. You can include furniture only if it is consistently included in MRc3 - MRc7.

% of Reused Materials	Points
5%	1
10%	2

Credit Path for CS:

A total of <u>5%</u> or more (based on <u>cost</u>) of the materials that you use should be <u>reused, refurbished or salvaged</u> materials. The approved materials can be from off-site sources.

Submittals:
1) A list of salvaged and reused materials and their related <u>costs</u>. Use construction costs for materials per <u>CSI MasterFormat™ 2004 Edition</u> Divisions 03-10, 31 (section 31.60.00 Foundations) and 32 (section 32.10.00 Paving, 32.30.00 Site Improvements, and 32.90.00 Planting).
OR
2) A list of <u>actual materials costs</u>, not including equipment and labor

Synergies:

MRc3 is related to the following credits:
- MRc1: Building Reuse
- MRc2: Construction Waste Management

<u>Remanufactured</u> materials cannot count towards MRc3 because they are NOT considered reuse materials, but they can contribute to the following credits:
- MRc2: Construction Waste Management
- MRc4: Recycled Content

Material costs for MRc3 shall be consistent with those used in the following credits:
- MRc4: Recycled Content
- MRc5: Regional Materials
- MRc6: Rapidly Renewable Materials

Possible Strategies and Technologies:
1) Find potential material suppliers and take advantage of every opportunity to use reused materials.
2) Try to reuse building materials like doors and frames, columns and beams, paneling, flooring, furniture and cabinetry, decorative items and bricks, etc.
3) <u>Exclude</u> repaired and recycled materials, <u>electrical, mechanical and plumbing items,</u> and <u>elevators and equipment or other specialty items</u> in your calculation.
4) The material cost is typically <u>45%</u> of the total cost of the building.

Extra Credit (Exemplary Performance):

You can get one innovation point for MRc3 if <u>15%</u> (for **NC and Schools**), OR <u>10%</u> (for **CS**) of the building materials are <u>reused, refurbished or salvaged</u>.

Project Phase: Construction Documents

<u>**LEED Submittal Phase:**</u> <u>Construction</u>

Related Code or Standard: None

Responsible Party: <u>Contractor and Architect</u>

MRc4: Recycled Content: (post-consumer + ½ pre-consumer) (1-2 points for NC, and Schools and CS)

Purpose:
To encourage the use of recycled materials in buildings, and reduce demand on virgin materials and the related environmental impact.

Credit Path for NC, and Schools:
At least 10% or 20% (based on cost) of the total value of the materials you use in the project should equal the sum of post-consumer plus ½ of the pre-consumer content.

% of Recycled Content	Points
10%	1
20%	2

You can use the weight of the recycled material divided by the total weight of the assembly to obtain the percentage (or fraction), and then use the percentage multiplied by the cost of the assembly to get the value of the recycled materials.

You should not include electrical, mechanical or plumbing items, elevator equipment or other specialty items in you calculation. You can only include permanently installed materials for your job. You can include furniture only if it is consistently included in MRc3 - MRc7.

You can define recycled materials per the International Organization of Standards document, *ISO 14021-Environmental labels and declarations- Self-declared environmental claims (Type II environmental labeling).*

Post-consumer materials refer to waste created by institutional, industrial or commercial facilities or households that can no longer be used for its intended purpose.

Pre-consumer materials refer to waste diverted from the waste flow in the process of manufacturing. You cannot include materials (like regrind, rework, etc.) that can be reclaimed within the same process that they were generated.

Submittals:
1) A list of actual materials costs, not including equipment and labor, for materials per CSI MasterFormat™ 2004 Edition Divisions 03-10, 31 (section 31.60.00 Foundations) and 32 (section 32.10.00 Paving, 32.30.00 Site Improvements, and 32.90.00 Planting), including Division 12 is optional
2) A list of manufacturers' names, product names, costs, and percentage of pre- and post-consumer content
3) Manufacturers' letters or cutsheets indicating the listed products' recycled content

Synergies:

MRc4 is related to the following credits:
- MRc2: Construction Waste Management
- MRc3: Materials Reuse

Look for recycled materials that are remanufactured locally and use local waste products for potential synergies with MRc5, Regional Materials.

Some recycled-content materials like synthetic products (rubber, plastic and polyester) may have problematic emissions. Make sure you consider coordinating all recycled-content materials with IEQc4, Low-Emitting Materials

Material costs for MRc4 shall be consistent with those used in the following credits:
- MRc3: Materials Reuse
- MRc5: Regional Materials
- MRc6: Rapidly Renewable Materials

Possible Strategies and Technologies:
1) Set a goal for recycled content, find potential material suppliers, and take advantage of every opportunity to recycle materials.
2) Make sure the recycled materials are installed.
3) Consider environmental, performance, and economic factors when choosing materials.
4) For **CS** projects, you can try to recycle building envelope materials and major structural elements because interior construction is NOT part of your scope of work.

Extra Credit (Exemplary Performance): You can get one innovation point for MRc4 if 30% of the materials are <u>recycled</u>.

Project Phase: Schematic Design

<u>**LEED Submittal Phase:**</u> <u>Construction</u>

Related Code or Standard:
<u>International Organization for Standardization (ISO)</u>: International Standard ISO 14021-1999, Environmental Labels and Declarations-Self-Declared Environmental Claims (Type II Environmental Labeling)

Responsible Party: <u>Contractor and Architect</u>

MRc5: Regional Materials (<u>1-2</u> points for NC, Schools, and CS)

Purpose:
To encourage the use of materials that are manufactured or extracted within the region, and reduce the environmental impact generated from transportation.

Credit Path:
At least <u>10%</u> or <u>20%</u> (based on <u>cost</u>) of the total value of the building products or materials that you use

in the project should be <u>recovered, harvested, extracted or manufactured</u> within <u>500 miles</u> of the job site. If only portion of a building product or material is generated locally, you should only include that portion (by percentage of weight) for this credit.

% of Regional Materials	Points
10%	1
20%	2

Do <u>not</u> include electrical, mechanical and plumbing items, and elevators and equipment or other specialty items in you calculation. You can only include <u>permanently</u> installed materials for your job. You can include furniture only if it is consistently included in MRc3 - MRc7.

Submittals:
1) A list of <u>actual materials costs</u>, not including equipment and labor, for materials per <u>CSI MasterFormat™ 2004 Edition</u> Divisions 03-10, 31 (section 31.60.00 Foundations) and 32 (section 32.10.00 Paving, 32.30.00 Site Improvements, and 32.90.00 Planting), including Division 12 is optional
2) A <u>list</u> of product purchases extracted, harvested, or manufactured regionally.
3) Paperwork showing manufacturers' names, product names, costs, <u>distance</u> between the extraction site and the project, and distance between the manufacturer and the project
4) Manufacturers' <u>letters or cutsheets</u> indicating the materials' origin and manufacture within a <u>500 miles radius</u> of the site

Synergies:
MRc5 may affect the following credits:
- MRc3: Materials Reuse
- MRc4: Recycled Content
- MRc6: Rapidly Renewable Materials

Material costs for MRc5 shall be consistent with those used in the following credits:
- MRc3: Materials Reuse
- MRc4: Recycled Content
- MRc6: Rapidly Renewable Materials

Possible Strategies and Technologies:
1) Set a goal for local materials, find potential material suppliers and take advantage of every opportunity to use local materials.
2) Make sure the local materials are installed.
3) Include environmental, performance, and economic factors when choosing materials.

Extra Credit (Exemplary Performance): You can get one innovation point for MRc5 if <u>30%</u> of the materials are from <u>local</u> sources.

Project Phase: Schematic Design

LEED Submittal Phase: <u>Construction</u>

Related Code or Standard: None

Responsible Party: <u>Contractor and Architect</u>
MRc6: Rapidly Renewable Materials (<u>1</u> point for NC and Schools. <u>Not Applicable</u> for CS)

Purpose:
To use rapidly renewable materials to replace long-cycle renewable and limited raw materials.

Credit Path for NC and Schools:
At least <u>2.5%</u> (based on <u>cost</u>) of the building products' and materials' total value for your job should be rapidly renewable materials (produced from plants that can be generated in a <u>10-year</u> cycle or less).

Submittals:
1) A list of <u>actual materials costs</u>, not including equipment and labor, for materials per <u>CSI MasterFormat™ 2004 Edition</u> Divisions 03-10, 31 (section 31.60.00 Foundations) and 32 (section 32.10.00 Paving, 32.30.00 Site Improvements, and 32.90.00 Planting), including Division 12 is optional
2) A <u>list</u> of rapidly renewable product purchases
3) Paperwork showing the manufacturers' names, product names, costs, percentage of rapidly materials (by <u>weight</u>) in each product, and compliance value
4) Manufacturers' <u>letters or cutsheets</u> confirming rapidly materials

Synergies:
Rapidly renewable materials may come from a distant location and may affect the following credits:
- MRc5: Regional Materials

Specify materials that do not release irritating or harmful chemicals like VOCs from solvent and paints. Coordinate MRc6 with the following credit:
- IEQc4: Low-Emitting Materials

Material costs for MRc6 shall be consistent with those used in the following credits:
- MRc3: Materials Reuse
- MRc4: Recycled Content
- MRc5: Regional Materials

Possible Strategies and Technologies:
1) Set a goal for rapidly renewable materials, find potential material suppliers, and take advantage of every opportunity to use rapidly renewable materials.
2) Make sure the rapidly renewable materials are installed.
3) Rapidly renewable materials include <u>wool, bamboo, cotton, wheatboard, cork, strawboard, cotton insulation, agrifiber, linoleum,</u> etc.

Extra Credit (Exemplary Performance): You can get one innovation point for MRc6 if <u>5%</u> of the materials are rapidly renewable materials.

Project Phase: Schematic Design

LEED Submittal Phase: <u>Construction</u>

Related Code or Standard: None

Responsible Party: Contractor and Architect

MRc7 for NC, Schools, and MRc6 for CS: Certified Wood (1 point for NC, Schools, and CS)

Purpose:
To encourage forest management that is environmentally friendly, to ban the use of pesticides, and to protect the health and safety of forest workers.

Credit Path for NC, Schools, and CS:
At least 50% (based on cost) of wood building components should be certified per **Forest Stewardship Council's (FSC)** principles and criteria. These wood components include general dimensional framing, structural framing, sub-flooring, flooring, finishes, wood doors, etc.

Only include permanently installed materials. You can include furniture only if it is consistently included in MRc3 - MRc7.

Exception to this rule: You can include purchased wood products for temporary use like bracing, formwork, scaffolding, guardrail and sidewalk protection, etc. If you decide to include these wood products, you need to include ALL such materials. If you purchase them for multiple projects, you can ONLY include them for one project of your choice.

Submittals:
1) A list identifying the certified wood in each purchase
2) Record of certified wood purchase and related **Chain of Custody (COC)** paperwork
3) Vendor invoices for certified wood products

Synergies:
Materials used for MRc7 may also qualify for:
- IEQc4.4: Low-Emitting Materials: Composite Wood and Agrifiber Products
- MRc5: Regional Materials

Possible Strategies and Technologies:
1) Set a goal for FSC certified wood products, find potential material suppliers, and take advantage of every opportunity to use rapidly renewable materials.
2) Make sure the FSC certified wood products are installed and quantified in terms of percentage.
3) There are two kinds of certification: a. Chain of Custody (COC) for the suppliers; b. All trades' forest management.
4) The suppliers need to get a certification number, but the contractor does NOT need to get one.
5) If you need to get a certification number for part of the product, then you need to get certification for the whole product.

Extra Credit (Exemplary Performance):

You can get one innovation point if <u>95%</u> of the <u>wood</u> building components are certified per **Forest Stewardship Council's (FSC)** Principles and Criteria.

Project Phase: Schematic Design

<u>LEED Submittal</u> Phase: <u>Construction</u>

Related Code or Standard:
<u>Forest Stewardship Council's (FSC)</u> principles and criteria

Responsible Party: <u>Contractor and Architect</u>

MR Summary and Mnemonics:

MR Credit Name	Extra Credit	Responsible Party
MRp1: Storage and Collection of Recyclables (Required for NC, Schools and CS)	0	Architect, Owner and tenant
*MRc1.1 for NC & Schools, *MRc1 for CS: Building Reuse: Maintain Existing Walls, Floors and Roof (1-3 points for NC, 1-2 points Schools, and 1-5 points CS)	For CS projects, you can get one ID extra point for reusing 95% or more of the building.	Architect, Contractor, and Owner
*MRc1.2 for NC & Schools: Building Reuse: Maintain Interior Nonstructural Elements (1 point for NC and Schools. N/A to CS)	0	Architect, Contractor, and Owner
MRc2: Construction Waste Management (1-2 points for NC, and Schools and CS)	1 extra ID point for MRc2 by salvaging and/or recycling a minimum of 95% of non-hazardous demolition and construction debris.	Contractor
MRc3: Materials Reuse (1-2 points for NC and Schools, and 1 point for CS)	1 extra ID point for MRc3 if 15% (for NC and Schools), OR 10% (for CS) of the building materials are reused, refurbished, or salvaged.	Contractor and Architect
MRc4: Recycled Content: (post-consumer + ½ pre-consumer) (1-2 points for NC, and Schools and CS)	1 extra ID point for MRc4 if 30% of the materials are recycled.	Contractor and Architect
MRc5: Regional Materials (1-2 points for NC, and Schools and CS)	1 extra ID point for MRc5 if 30% of the materials are from local sources.	Contractor and Architect
*MRc6: Rapidly Renewable Materials (1 point for NC and Schools. N/A for CS)	1 extra ID point for MRc6 if 5% of the materials are rapidly renewable materials.	Contractor and Architect
*MRc7 for NC and Schools, and *MRc6 for CS: Certified Wood (1 point for NC, Schools, and CS)	1 extra ID point if 95% of the wood building components are certified per Forest Stewardship Council's (FSC) principles and criteria.	Contractor and Architect

Mnemonics:

SBC

MR Regan Ray Carter (See bold and underlined letters in credit names on table above also). Of course, you can create your own **mnemonics** to help you memorize them.

Note: * indicates prerequisite or credit NOT applicable to all LEED rating systems. Detailed discussions have been omitted since the information is unlikely to be tested on the LEED Green Associate Exam.

E. Indoor Environmental Quality (IEQ)

The average American spends about 90% of his time indoors, so indoor environmental quality is very important for quality of life, well-being, and productivity.

Overall purpose:
1) Design and install proper systems to achieve a quality indoor environment
2) Reduce, manage, and eliminate contaminants
3) Minimum IAQ Performance
4) Environmental Tobacco Smoke (ETS) Control
5) Outdoor Air Delivery Monitoring
6) Increased Ventilation
7) Construction IAQ Management Plan
8) Low-Emitting Materials
9) Indoor Chemical and Pollutant Source Control
10) Controllability of Systems
11) Thermal Comfort
12) Daylight and Views

Mnemonics:
I Call Mike Evans.
Oh! Ian Catches LIST D (See underlined letters in credit names on table above also.)

Core concepts:
1) Improve indoor air quality
   • Improve building ventilation
   • Choose proper materials
   • Reduce, manage, and eliminate contaminants
   • Advocate green construction practices and green building operation
2) Improve indoor environmental quality
   • Thermal comfort control
   • Daylight and views
   • Considering acoustics

Recognition, regulation and incentives:
1) Guidance is available from private and public sector organizations such as:
   • EPA
   • Department of Labor
   • ASHRAE

Overall strategies and technologies:
1) Choose low-emitting material, interior finishes, furniture, etc.
2) Reduce, manage, and eliminate contaminants like certain cleaning products, tobacco smoke, and radon.
3) Advocate green construction practices like **Best Management Practices (BMPs)**, outdoor air introduction, green cleaning, and proper handling of exhaust systems.
4) Thermal comfort control: user control and feedback, operations and maintenance/ management.
5) Provide daylight/views: north-facing skylight, interior light (reflecting) shelf, interior and exterior permanent shading devices, automatic photocell-based control and high performance glazing to maximize daylight for interior spaces while avoiding high-contrast conditions.
6) Use light fixture with sensors and dimming controls.

Specific Technical Information:

IEQp1: Minimum IAQ Performance (Required for NC, Schools, and CS)

Purpose:
To improve indoor air quality and the well being and comfort of building occupants by establishing minimum indoor air quality (IAQ) performance.

Credit Paths for NC, Schools, and CS:
1) **Active (Mechanical)** Ventilation: Your building should meet Section 4 through 7 of ASHRAE 62.1-2007, Ventilation for Acceptable Indoor Air Quality (with errata but without addenda). Your mechanical ventilation systems shall follow Ventilation Rate Procedure or local codes, if they are more restrictive.

OR

2) **Passive (Natural)** Ventilation: Your building should follow ASHRAE 62.1-2007 paragraph 5.1 (with errata but without addenda), if it is naturally ventilated.

Additional Requirements for CS: CS ventilation systems should meet the anticipated tenant requirements.

Regardless of the ventilation mode, your building should follow ASHRAE 62.1-2007, Chapter 6 (with errata but without addenda).

Submittals:
1) Calculations to confirm compliance with ASHRAE 62.1-2007
2) For **CS** projects, a description of future tenants, expected uses and spaces types

Synergies:
You can dilute contaminant concentration and solve some of the IAQ problems, but this measure

may increase energy use and affect indoor thermal comfort. There are ways to minimize energy loss and improve IAQ as described in the following prerequisite and credits:

- EAp1: <u>Fundamental</u> Commissioning of the Building Energy Systems

- EAc3: <u>Enhanced</u> Commissioning
- EAc5: <u>Measurement and Verification</u>

Existing site contamination, heavy traffic and dense neighborhoods may have a negative impact on the quality of outside air for ventilation. See credits below:

- SSc3: Brownfield Redevelopment
- SSc4: Alternative Transportation

Some materials may have negative effects on IAQ. Try to specify furnishings and materials free of irritating and harmful chemicals like VOCs from solvents and paints. See the following prerequisite and credits:

- IEQp2: Environmental <u>Tobacco</u> Smoke (ETS) Control

- IEQc4: <u>Low-Emitting</u> Materials
- IEQc5: Indoor Chemical and Pollutant <u>Source</u> Control

Possible Strategies and Technologies:
1) Design your ventilation system to meet or exceed ASHRAE requirements. Optimize and balance occupant health with energy efficiency.
2) See the ASHRAE 62 User's Manual for more information.
3) For **Passive (Natural)** Ventilation, the openings for ventilation shall be at least 4% of the floor area.

Extra Credit (Exemplary Performance): None

Project Phase: Schematic Design

<u>LEED Submittal</u> Phase: <u>Design</u>

Related Code or Standard:
<u>ANSI/ASHRAE 62.1-2007 (with errata but without addenda)</u>

Responsible Party: <u>MEP Engineer</u>

IEQp2: Environmental Tobacco Smoke (ETS) Control (Required for NC, Schools, and CS)

Purpose:
For NC and CS
To <u>minimize</u> the impact of Environmental Tobacco Smoke (ETS) on ventilation air distribution systems, indoor surface and occupants.

For Schools

To eliminate the impact of Environmental Tobacco Smoke (ETS) on ventilation air distribution systems, indoor surface and occupants.

Credit Paths for NC and CS:

1st Path:

    a. Place any designated exterior smoking area a minimum of 25 feet away from operable windows, entries, and outdoor air intakes. Use signage to designate smoking areas, and ban smoking in certain areas, or ban smoking on the entire property.

    b. Ban indoor smoking.

OR

2nd Path (For **nonresidential** projects):

    a. Place any designated exterior smoking area a minimum of 25 feet away from operable windows, entries, and outdoor air intakes. Use signage to designate smoking areas, and ban smoking in certain areas, or ban smoking on the entire property.

    b. Ban indoor smoking, except in designated smoking areas/rooms.

    c. Place smoking rooms at a location to effectively contain, capture and remove ETS. The smoking room should be exhausted directly to the outdoors and enclosed with deck-to-deck impermeable partitions. The ETS-containing air shall never be re-circulated to non-smoking areas. When the smoking room doors are closed, the smoking room should have negative pressure of a minimum of 1 Pa (0.004 inches of water gauge), and an average of 5 Pa (0.02 inches of water gauge).

    d. You need to have a 15-minute measurement conducted with the smoking room doors closed. You should have 1 measurement of at least every 10 seconds of the pressure between the smoking rooms, each adjacent vertical chase, and each adjacent area. The tests should be done for the worst case conditions.

3rd Path (For **residential** and hospitality projects):

    a. Place any designated exterior smoking area a minimum of 25 feet away from operable windows, entries and outdoor air intakes to common areas. Use signage to designate smoking areas, and ban smoking in certain areas, or ban smoking on the entire property.

    b. Ban smoking in building's common areas.

    c. Sealing vertical chases, penetrations in floors, walls, and ceilings to minimize ETS transfer between units.

    d. For residential units, install weather-striping on all operable windows, exterior doors, and doors leading to common hallways.

    e. The residential units adjacent to the common pathways do not need to have weather-striping if the common hallways are pressurized per Path 2 above.

    f. The sealing of residential units shall pass a blower door test per ANSI/ASTM-E779-03, **Standard Test Method** for Determining Air Leakage Rate by Fan Pressure, and the **Progressive Sampling Method** per Chapter 4 (Compliance Through Quality Construction) of the Residential Manual for Compliance with California's 2001 Energy Efficiency Standards (www.energy.ca.gov/). Residential units should have less than 1.25 square inches leakage area per 100 square feet of enclosed area (the total of floor, wall and ceiling areas).

Credit Paths for Schools

    a. Place any designated exterior smoking area a minimum of 25 feet away from operable windows, entries, and outdoor air intakes. Use <u>signage</u> to designate smoking areas, and ban smoking in certain areas or ban smoking on the entire property.

b. <u>Ban</u> indoor smoking.

Submittals:
1) An ETS <u>policy</u> describing areas of non-smoking
2) Renderings, site plans or other <u>documentation</u> to show <u>how</u> the smoking policy will be implemented
3) A record of <u>testing data</u> for interior smoking rooms to show that there is <u>no cross contamination</u>

Synergies:
A separate ventilation system for smoking areas adds cost to commissioning, energy, and M&V efforts. See the following prerequisite and credits:
- EAp1: <u>Fundamental</u> Commissioning of the Building Energy Systems
- EAc1: <u>Optimize</u> Energy Performance
- EAc3: <u>Enhanced</u> Commissioning
- EAc5: <u>Measurement and Verification</u>

Both outdoor and indoor smoking affects IAQ performance. IEQp2 is related to the following prerequisite and credits:
- IEQp1: Minimum IAQ Performance
- IEQc1: Outdoor Air Delivery Monitoring
- IEQc2: Increased Ventilation

You may want to address smoking with other related air pollutants. See following credits:
- IEQc4: Low-Emitting Materials
- IEQc5: Indoor Chemical and Pollutant Source Control Products

Possible Strategies and Technologies:
1) For commercial buildings, effectively design and control ventilation of smoking rooms or ban smoking.
2) For residential projects, ban smoking in common areas, and design and construct building systems and envelope to minimize ETS transfer between the dwelling units and common areas.

Extra Credit (Exemplary Performance): None

Project Phase: Pre-Design

LEED Submittal Phase: <u>Design</u>

Related Code or Standard:
1) <u>ANSI/ASTM</u>-E779-03, Standard Test Method for Determining Air Leakage Rate by Fan Pressure
2) Residential Manual for Compliance with <u>California's 2001 Energy Efficiency Standards</u> (Low Rise Residential Buildings, Title 24 or T-24), Chapter 4

Responsible Party: <u>LEED AP</u>, Owner, and MEP Engineer

IEQp3: Minimum Acoustical Performance (For Schools only)

Note: Detailed discussions have been omitted since this prerequisite is for schools only, and is unlikely to be tested on the LEED Green Associate Exam.

IEQc1: Outdoor Air Delivery Monitoring (<u>1</u> Point for NC, Schools, and CS)

Purpose:
The intent is to improve occupant well-being and comfort by providing ways to monitor ventilation systems.

Credit Paths for NC, Schools, and CS:
Install <u>permanent</u> monitoring systems to check the performance of the ventilation system. Set an alarm to go off when the ventilation system performance (CO_2 levels or air flow values) deviates <u>10%</u> or more from the setpoint. The alarm should be audible or visual to the system operator.

1) For <u>Naturally</u> Ventilated Spaces

Monitor CO_2 concentrations in all spaces that are naturally ventilated. Place CO_2 monitoring equipment between <u>3' and 6'</u> above the floor in the rooms. You can use one CO_2 sensor for multiple spaces if you use passive stack(s) or other measures to induce air flow through these spaces simultaneously and equally without the help of occupants.

2) For <u>Mechanically</u> Ventilated Spaces

<u>Monitor CO_2</u> concentrations in all densely occupied spaces, i.e., spaces with a design occupant density <u>greater than or equal to 25 persons per 1000 s.f.</u> Place CO_2 monitoring equipment between 3' and 6' above the floor in the rooms.

For spaces with a design occupant density <u>less than 25 persons per 1,000 s.f.</u>, you should provide a <u>direct outdoor airflow monitoring/measurement device</u> that is able to check the minimum outdoor airflow rate with an accuracy rate of plus or minus <u>15%</u> of the minimum design outdoor airflow rate per ASHRAE 62.1-2007 (with errata but without addenda).

Submittals:
1) <u>Show CO_2 sensors and air flow monitors</u> on schematics, floor plans, elevations and mechanical schedules.
2) Use commissioning to monitor the <u>ventilation system energy usage</u>
3) Check <u>alarm system settings</u> per ANSI/ASHRAE 62.1-2007 for mechanical ventilation systems
4) Adjust all <u>automated building systems</u> per manufacturer's recommendations. Perform routine check for alarm systems

Synergies:
EAc1 can alert building users about potential IAQ problems and create opportunities for performance trending. Monitoring capacity enables robust M&V and provides information for

commissioning, ensures consistent IAQ, and maximizes energy performance. See the following prerequisite and credits:

* IEQc2: Increased Ventilation

* EAp1: <u>Fundamental</u> Commissioning of the Building Energy Systems
* EAc3: <u>Enhanced</u> Commissioning
* EAc5: Measurement and Verification

You can increase the ventilation rates to mitigate indoor CO_2, but outdoor air quality depends on the CO_2 in the outside air. Heavy traffic, dense neighborhoods, and existing site contamination can increase CO_2 levels and affect the outdoor air quality. Bicycle corridors, public transportation and other transportation amenities can reduce the number of solo drivers and lower CO_2 levels. See:

* SSc4: Alternative Transportation

Possible Strategies and Technologies:
1) Install airflow and carbon dioxide monitoring equipment, and then transfer the information to the <u>Building Automation System (BAS)</u> and/or HVAC system to trigger corrective action.
2) If you can<u>not</u> do these, then at least use the monitoring equipment to trigger alarms to alert the occupants or operators of the poor outdoor air delivery.

Extra Credit (Exemplary Performance): None

Project Phase: Schematic Design

<u>LEED Submittal</u> Phase: <u>Design</u>

Related Code or Standard:
1) <u>ANSI/ASHRAE 62.1-2007</u>: Ventilation for Acceptable Indoor Air Quality

Responsible Party: <u>MEP Engineer</u>

IEQc2: Increased Ventilation (<u>1</u> Point for NC, Schools, and CS)

Purpose:
To provide more outdoor air ventilation to improve indoor air quality, occupants' well-being, comfort, and productivity.

Credit Paths for NC, Schools and CS:
1) **For <u>Naturally</u> Ventilated Spaces**
   Study and follow the flow diagram process in Figure 2.8 of the Chartered Institution of Building Service Engineers (**CIBSE**) Applications Manual 10: 2005, natural ventilation in non-domestic buildings, and determine natural ventilation is an effective design for your job.

AND

Use calculations and diagrams to show that your system meets the above mentioned CIBSE requirements:

1ˢᵗ Option: CIBSE Applications Manual 10: 2005, Natural Ventilation in Non-Domestic Buildings

OR

2ⁿᵈ Option: CIBSE AM 13:2000, Mixed Mode Ventilation

OR

2) Use an analytic, multi-zone, macroscopic model to show that airflow in each room will effectively naturally ventilate and meet the minimum ventilation rates per <u>ASHRAE 62.1-2007</u> Chapter 6 <u>(with errata but without addenda)</u>, for a minimum of <u>90%</u> of the occupied spaces.

OR

3) **For <u>Mechanically</u> Ventilated Spaces**

Increase outdoor air ventilation rates to breathing zones of all occupied spaces by a minimum of <u>30%</u> above the minimum rates per <u>ASHRAE Standards 62.1-2007</u>. See EQp1 for more information.

Note:

For ALL options, breathing zone means space <u>2'</u> from exterior walls and <u>3' to 6'</u> above finished floor (AFF). Use <u>0.30 CFM per s.f.</u> for outside air.

Submittals:
1) <u>Size mechanical equipment</u> to match increase ventilation rates
2) Keep plans and other proper <u>documentation of open areas</u> within the project for naturally ventilated projects

Synergies:

Ventilation, especially mechanical ventilation, requires commissioning, M&V, and building energy performance. A permanent ventilation performance monitoring system can help to achieve and maintain increased ventilation. See the following related prerequisite and credits:
- EAp1: <u>Fundamental</u> Commissioning of the Building Energy Systems
- EAp2: <u>Minimum</u> Energy Performance
- EAc1: <u>Optimize</u> Energy Performance
- EAc3: <u>Enhanced</u> Commissioning
- EAc5: <u>Measurement and Verification</u>
- IEQc1: <u>Outdoor</u> Air Delivery Monitoring

Possible Strategies and Technologies:

Follow the eight design steps in <u>Carbon Trust "Good Practice Guide 237"</u> for <u>naturally</u> ventilated spaces:
1) Develop <u>design</u> requirements

2) Plan airflow paths
3) Identify building uses and features that might require special attention
4) Determine ventilation requirements
5) Estimate external driving pressure
6) Select types of ventilation devices
7) Size ventilation devices
8) Analyze the design

You can use LoopDA, Natural Ventilation Sizing Tool, and NIST's CONTAM, Multizone Modeling Software to show airflows in each room analytically.

Use heat recovery if possible to reduce energy consumption caused by higher ventilation rates for mechanically ventilated spaces.

Extra Credit (Exemplary Performance): None

Project Phase: Schematic Design

LEED Submittal Phase: Design

Related Code or Standard:
1) Chartered Institution of Building Service Engineers (CIBSE)
2) Carbon Trust "Good Practice Guide 237"
3) ASHRAE Standards 62.1-2007

Responsible Party: Contractor and MEP Engineer

IEQc3.1 for NC and Schools, and IEQc3 for CS: Construction IAQ Management Plan: During Construction (1 Point for NC, Schools, and CS)

Purpose:
To sustain the well-being and comfort of building occupants and construction workers by reducing air pollution caused by the renovation/construction process.

Credit Paths for NC, Schools, and CS:
Create and implement an IAQ (Indoor Air Quality) Management Plan for the pre-occupant and construction phases:
1) Use filtration media with a Minimum Efficiency Reporting Value (MERV) of **8** at each return air grille per ASHRAE 52.2-1999 if you use a permanently installed air handler during construction. Change all filtration media immediately before occupancy.
2) Protect installed or stored on-site absorptive materials from moisture damage.
3) At minimum, meet the Control Measures of the Sheet Metal and Air Conditioning National Contractors Association (SMACNA) IAQ Guideline for Occupied Buildings under Construction, 2nd Edition, November 2007, Chapter 3, during construction.

Additional Requirements for Schools:

Once the building is enclosed, ban smoking within 25 feet of building entrances and inside the building.

Submittals:
1) A written construction IAQ management plan for demolition and construction
2) A detailed photo log of construction IAQ management plan practices used during construction

Synergies:
See the following related prerequisite and credits:
- IEQc3.2: Construction IAQ Management Plan: Before Occupancy
- IEQc4: Low-Emitting Materials
- IEQc5: Indoor Chemical and Pollutant Source Control

Possible Strategies and Technologies:
1) Interrupt contamination pathways, control pollutant sources and protect the HVAC system by creating and implementing an IAQ management plan.
2) Properly sequence your material installation to avoid contamination of carpeting, ceiling tiles and gypsum wallboard, insulation, and other absorptive materials. Coordinate with IEQc3.2 and IEQc5; select the proper schedules and specifications for filtration media.
3) Try not to use permanently installed air handlers for temporary cooling/heating during construction. See related USGBC reference guide for more information.

Extra Credit (Exemplary Performance): None

Project Phase: Construction Administration

LEED Submittal Phase: Construction

Related Code or Standard:
1) ASHRAE 52.2-1999
2) Sheet Metal and Air Conditioning National Contractors Association (SMACNA) IAQ Guideline for Occupied Buildings under Construction, 2nd Edition, November 2007, Chapter 3.

Responsible Party: Contractor and MEP Engineer

IEQc3.2: Construction IAQ Management Plan: Before Occupancy (1 Point for NC and Schools, NOT applicable for CS)

Purpose:
To sustain the well-being and comfort of building occupants and construction workers by reducing air pollution caused by the renovation/construction process.

Credit Paths for NC and Schools:
Create and implement an IAQ (Indoor Air Quality) Management Plan after the installation of all finishes and the cleaning of the entire building but before occupancy:

1st Path (Flush-out):

After construction and all interior finishes are completed but before occupancy, do a flush-out by keeping the relative humidity less than or equal to <u>60%</u>, and a minimum of <u>60</u> degrees Fahrenheit internal temperature, and provide a total air volume of <u>14,000 cubic feet (c.f.)</u> of outdoor air per s.f. of floor area.

OR

If you want to use a space before flush-out, you should supply at least <u>3,500 c.f.</u> of outdoor air per s.f. of floor area before you can use the space. Once you occupy the space, you need to ventilate it at least at a rate of <u>0.30cfm/s.f.</u> of outside air or meet the minimum requirement per EQp1 if it is more restrictive. You should ventilate the space for at least <u>three hours</u> before you use the space and continue to ventilate it while you use the space. You need to maintain these until you supply a total of <u>14,000 c.f./s.f.</u> of outside air.

Note: You need to start the flush-out AFTER completion of cleaning, the punch list, and final test of the control systems.

OR

2nd Path (Air Testing):

a. Per the U.S. <u>Environmental Protection Agency (EPA)</u> Compendium of Methods for the Determination of Air Pollutants in Indoor Air, do a baseline IAQ test after construction, but before occupancy. See related USGBC Reference Guide for more information.
   Also make sure that your spaces have not exceeded the maximum contaminant concentrations listed per the related USGBC reference guide.
b. If any of the limits above is exceeded at a sampling point, you need to <u>do additional flush-outs until</u> a retest showing the above limit is met. If you need to retest non-complying areas, take samples from the same locations as in the first test.

c. Conduct air sampling tests per the following criteria:
1) Before occupancy, simulate the normal occupancy conditions, i.e., do the test during <u>normal</u> hours, start the ventilation at normal hours, and run it at <u>minimum</u> outside air flow rate for the occupied mode for the entire length of the testing.
2) All permanent interior finishes (carpet and acoustic ceiling tiles, paint, doors, millwork, etc.) should have been <u>completed</u>. You are not required to complete installation of movable furnishings.
3) The number of ventilation systems and size of the building dictate the number of sampling locations. You should have at least <u>one</u> sampling point for each continuous floor area or for every <u>25,000 s.f.</u> if a larger number is required. Also include the areas with the worst ventilation and with the best ventilation.
4) Take the samples between <u>3' to 6'</u> high (the breathing zone of a person) and for at least <u>4 hours</u>.

Submittals:
1) A <u>written</u> construction IAQ management <u>plan</u>
2) <u>Documentation</u> for occupancy, dates, outdoor air monitoring rates, humidity, internal temperature, and special considerations for <u>flush-out procedure</u>
3) A copy of the <u>testing report</u>, confirmation of the correct <u>unit</u> of measurement used and inclusion of all <u>contaminants</u> for <u>IAQ testing</u>

Synergies:

The same construction IAQ management plan often includes best practices during construction and after construction but before occupancy. See:

- IEQc3.1: Construction IAQ Management Plan: During Construction

Building external moisture barrier and filtration materials can affect IAQ and related testing. See the following related prerequisite and credits:

- IEQc4: Low-Emitting Materials
- IEQc5: Indoor Chemical and Pollutant <u>Source</u> Control

You can introduce outdoor air and dilute indoor air contaminants. See the following related prerequisite and credits:

- IEQp1: Minimum IAQ Performance
- IEQc2: Increased Ventilation

Possible Strategies and Technologies:

1) Do a flush-out if you do not need to use the building right away.
2) If you do need to use the building right away, you can do tests for air contaminant levels, but the tests can be more expensive than a flush-out. You should coordinate with IEQc3.1 and IEQc5 for filtration media.

Extra Credit (Exemplary Performance): None

Project Phase: Construction Administration

<u>LEED Submittal</u> Phase: <u>Construction</u>

Related Code or Standard:
U.S. <u>Environmental Protection Agency (EPA)</u> Compendium of Methods for the Determination of Air Pollutants in Indoor Air

Responsible Party: <u>Contractor and MEP Engineer</u>

IEQc4.1: Low-Emitting Materials: Adhesive and Sealants (<u>1</u> Point for NC, Schools and CS)

Purpose:
To reduce the number of harmful, irritating and/or odorous indoor air contaminants, and to provide comfort and protect the well-being of building users and workers who install the materials.

Credit Path for NC and CS:
Use building interior sealants and adhesives (applied on-site and placed inside of the weatherproofing system) per the following standards:

a. Sealants, Sealant Primers and Adhesive: <u>SCAQMD (South Coast Air Quality Management District) Rule #1168</u>, VOC limits effective 7/1/2005.

See table in related USGBC reference guide for VOC limits with an effective date of 7/1/2005, and rule amendment dated 1/7/2005.

Note: VOC <u>means "Volatile Organic Compound."</u> It is a major contributing factor to ozone depletion, a common air pollutant which has been proven to be a public health hazard.

b. Aerosol Adhesive: <u>Green Seal Standard</u> for Commercial Adhesives <u>GS-36</u> requirements with an effective date of 10/19/2000. See related USGBC reference guide for the table of VOC limits.

Credit Path for Schools:
Use building interior sealants and adhesives (applied on-site and placed inside of the weatherproofing system) per the product and testing requirements of the <u>California Department of Health Service Standard Practice</u> for the Testing of Volatile Organic Emissions from Various Sources Using Small-Scale Environmental Chambers, including 2004 Addenda.

Submittals:
1) A <u>list</u> of interior sealants, sealant primers, and aerosol adhesive used, including product name, manufacturer's name, specific VOC data (g/L, less water), and the related allowable VOC from the reference standard
2) If you use VOC budget approach, track the amount used

Synergies:
See the following related prerequisite and credits:
- IEQp2: Environmental <u>Tobacco</u> Smoke (ETS) Control
- IEQp3: Minimum Acoustical Performance (for Schools only)

- IEQc3.1: Construction IAQ Management Plan: During Construction
- IEQc3.2: Construction IAQ Management Plan: Before Occupancy

- IEQc4.2: Low-Emitting Materials: Paints and Coatings
- IEQc4.3: Low-Emitting Materials: Carpet Systems
- IEQc4.4: Low-Emitting Materials: Composite Wood and Agrifiber Products
- IEQc4.5 Low-emitting Materials-Furniture and Furnishings Systems (for Schools only)
- IEQc4.6 Low-emitting Materials-Ceiling and Wall Systems (for Schools only)

- IEQc5: Indoor Chemical and Pollutant <u>Source</u> Control

- IEQc9: Enhanced Acoustical Performance (for schools only)

Possible Strategies and Technologies:
1) Make sure that you list the VOC limits, or at least refer to related USGBC Reference Guide for VOC limits in each specification section.
2) Request cutsheets, MSD and certification as part of the submittals and review them for VOC limit compliance.
3) Mention VOC limits in the pre-bid meeting and bid-award meeting.

Extra Credit (Exemplary Performance): None

Project Phase: Design Development

LEED Submittal Phase: Construction

Related Code or Standard:
1) SCAQMD (South Coast Air Quality Management District) Rule #1168, VOC limits, effective 1/7/2005
2) Green Seal Standard for Commercial Adhesives GS-36 requirements, effective 10/19/2000
3) California Department of Health Service Standard Practice for the Testing of Volatile Organic Emissions from Various Sources Using Small-Scale Environmental Chambers, including 2004 Addenda

Responsible Party: Contractor and Architect

IEQc4.2: Low-Emitting Materials: Paints and Coatings (1 Point for NC, Schools, and CS)

Purpose:
To reduce the number of harmful, irritating and/or odorous indoor air contaminants, and to provide comfort and protect the well-being of building users and workers who install the materials.

Credit Path for NC and CS:
Use building interior paints and coatings (applied on-site and placed inside of the weatherproofing system) per the following standards:

Architectural coatings, paints and primer used on interior ceilings and walls shall not exceed the VOC content limits per Green Seal Standard GS-11, Paints, 1st Edition, 5/20/1993.
> Flats: 50 g/L
> Non-Flats: 150 g/L

Anti-rust and anti-corrosive paints used on interior ferrous metal substrates shall not exceed the VOC content limits of 250 g/L per Green Seal Standard GS-03, Anti-Corrosive Paints, 2nd Edition, 1/7/1997.

Floor coatings, clear wood finishes, primers, stains and shellacs used for interior items shall not exceed the VOC content limits per SCAQMD (South Coast Air Quality Management District) Rule 1113, Architectural Coatings, rules with an effective date of 1/1/2004.
> Clear wood finishes: varnish 350 g/L; Lacquer 550 g/L
> Floor coatings: 100 g/L
> Sealers: waterproofing sealers 250 g/L; sanding sealers 275 g/L; all other sealers 200 g/L
> Shellacs: Clear 730 g/L; pigmented 550 g/L
> Stains: 250 g/L

Credit Path for Schools:
Use building interior sealants and adhesives (applied on-site and placed inside of the weatherproofing system) per the product and testing requirements of the <u>California Department of Health Service Standard Practice</u> for the Testing of Volatile Organic Emissions from Various Sources Using Small-Scale Environmental Chambers, including 2004 Addenda.

Submittals:
1) A <u>list</u> of interior coating products and paints used, including product names, manufacturers' name, specific VOC data (g/L, less water), and the related allowable VOC from the reference standard
2) If you use the VOC budget approach, track the <u>amount</u> used

Synergies:
Same list of prerequisites and credits as those for EAc4.1, except replace EAC4.2 with EAc4.1

Possible Strategies and Technologies: Same as IEQc4.1

Extra Credit (Exemplary Performance): None

Project Phase: Design Development

LEED Submittal Phase: <u>Construction</u>

Related Code or Standard:
1) <u>SCAQMD (South Coast Air Quality Management District) Rule 1113</u>, Architectural Coatings, rules with an effective date of 1/1/2004.
2) <u>Green Seal Standard GS-03 and GS-11</u>
3) <u>California Department of Health Service Standard Practice</u> for the Testing of Volatile Organic Emissions from Various Sources Using Small-Scale Environmental Chambers, including 2004 Addenda

Responsible Party: <u>Contractor and Architect</u>

IEQc4.3: Low-Emitting Materials: Carpet Systems (1 Point for NC, Schools, and CS)

Purpose:
To reduce the number of harmful, irritating and/or odorous indoor air contaminants, and to provide comfort and protect the well-being of the building users and the workers who install the materials.

Credit Path:
1) Credit Path for NC and CS:
   All flooring shall meet the following criteria as applicable:

- Make sure all <u>carpet</u> used in the interior of your building meets the product and testing requirements of the <u>Carpet and Rug Institute's Green Label Plus</u> program.

- Make sure all carpet <u>cushions</u> used in the interior of your building meets the requirements of the <u>Carpet and Rug Institute's Green Label</u> program.

- Make sure all carpet <u>adhesive</u> meets the VOC limit of <u>50 g/L</u> per IEQc4.1. The test results submitted for LEED certification shall not be more than <u>two years</u> old.

- Make sure all <u>hard</u> surface flooring (linoleum, vinyl, laminate flooring, wood flooring, ceramic flooring, rubber flooring and wall base) are certified to meet the <u>FloorScore™ standard</u> by a third party.

- Bamboo, cork, wood, and concrete floor finishes like stain and sealer finish must meet the requirements of <u>SCAQMD (South Coast Air Quality Management District) Rule 1113</u>, Architectural Coatings, rules with an effective date of 1/1/2004.

- Grout and setting adhesive shall meet the requirements of <u>SCAQMD (South Coast Air Quality Management District) Rule 1168</u> on VOC limits with an effective date of 7/1/2005, and rule amendment dated 1/7/2005.

OR

2) Credit Path for NC, Schools, and CS:

Ensure all the interior flooring elements in your building meet the product and testing requirements of the <u>California Department of Health Service Standard Practice</u> for the Testing of Volatile Organic Emissions from Various Sources Using Small-Scale Environmental Chambers, including 2004 Addenda.

Submittals:
1) A <u>list</u> of interior carpet adhesives, carpet and carpet cushions used, including VOC content of each
2) A <u>list</u> of tile setting adhesive, grout, finishes and other hard surface flooring product used, including VOC content of each tile setting grout and adhesive

Synergies:
Same list of prerequisites and credits as those for EAc4.1, except replace EAC4.3 with EAc4.1.

Possible Strategies and Technologies:
Make sure that you list the product certification and testing requirements in the plans or specifications. Choose products that meet the proper requirements and are tested by an independent and qualified lab or certified by the Green Label Plus program.

You can find information on the Green Label Plus program for carpets and the related VOC emission limit (microgram per square meter per hour), and sample collection and testing method by <u>Carpet and Rug Institute (CRI)</u> with the help of <u>California Department of Health Service (DHS)</u> and California's Sustainable Building Task Force in Section 9, Acceptable Emission Testing for Carpet, DHS Standard Practice CA/DHS/EHLB/R-174 with a date of 7/15/2004. Use the link below to go to DHS website and search for this document by it title:

http://www.dhs.ca.gov/

It is also published by the Collaborative for High Performance Schools (www.chps.net) as Section 01350 Section 9 (dated 2004).

Extra Credit (Exemplary Performance): None

Project Phase: Design Development

LEED Submittal Phase: Construction

Related Code or Standard:
1) Carpet and Rug Institute's Green Label Plus program
2) SCAQMD (South Coast Air Quality Management District) Rule 1168 on VOC limits with an effective date of 7/1/2005, and rule amendment dated 1/7/2005
3) SCAQMD (South Coast Air Quality Management District) Rule 1113, Architectural Coatings, rules with an effective date of 1/1/2004
4) California Department of Health Service Standard Practice for the Testing of Volatile Organic Emissions from Various Sources Using Small-Scale Environmental Chambers, including 2004 Addenda
5) State of California Standard 1350, Standard Practice for the Testing of Volatile Organic Emissions from Various Sources Using Small-Scale Environmental Chambers, Testing Criteria
6) FloorScore™ program

Responsible Party: Architect

IEQc4.4: Low-Emitting Materials: Composite Wood and Agrifiber Products (1 Point for NC, Schools, and CS)

Purpose:
To reduce the number of harmful, irritating and/or odorous indoor air contaminants, and to provide comfort and protect the well-being of the building users and workers who install the materials.

Credit Path for NC and CS:
Use agrifiber and composite wood products that contain NO added urea-formaldehyde resins on the building interior, i.e., inside of the weatherproofing system. Use laminated adhesives that contain no added urea-formaldehyde resins to fabricate shop-applied and on-site agrifiber and composite wood products.

Composite wood products include door cores, panel substrate, MDF (medium density fiberboard), particleboard, plywood, strawboard, and wheatboard. They do not include FF&E (fit-out, furniture, and equipment) since they are not base building elements.

Agrifiber wood products include cereal straw, coconut husk, sunflower husk, sugarcane, and agricultural pruning.

Credit Path for NC, Schools, and CS:

Ensure all <u>agrifiber</u> and <u>composite</u> wood products in your building meet the product and testing requirements of the <u>California Department of Health Service Standard Practice</u> for the Testing of Volatile Organic Emissions from Various Sources Using Small-Scale Environmental Chambers, including 2004 Addenda.

Submittals:

1) A <u>list</u> of interior agrifiber and composite wood product used, and confirmation that none of the product contain urea-formaldehyde

Synergies:

Same list of prerequisites and credits as those for EAc4.1, except replace EAC4.4 with EAc4.1

Possible Strategies and Technologies:

Make sure that you only list the agrifiber and composite wood products, field and shop assemblies, and <u>laminating adhesive</u> that contain <u>no</u> added urea-formaldehyde resins.

Extra Credit (Exemplary Performance): None

Project Phase: Design Development

<u>LEED Submittal</u> Phase: <u>Construction</u>

Related Code or Standard:

For NC and CS
None

For Schools
<u>California Department of Health Service Standard Practice</u> for the Testing of Volatile Organic Emissions from Various Sources Using Small-Scale Environmental Chambers, including 2004 Addenda

Responsible Party: <u>Contractor and Architect</u>

IEQc4.5   Low-emitting Materials: Furniture and Furnishings Systems (<u>1</u> Point for Schools Only)

Note: Detailed discussions have been omitted since this credit is for schools only, and is unlikely to be tested on the LEED Green Associate Exam.

IEQc4.6   Low-emitting Materials: Ceiling and Wall Systems (1 Point for Schools Only)

Note: Detailed discussions have been omitted since this credit is for schools only, and is unlikely to be tested on the LEED Green Associate Exam.

IEQc5: Indoor Chemical and Pollutant Source Control Products (1 Point for NC, Schools and CS)

Purpose:
To minimize building users' exposure to chemical pollutants and possibly dangerous particulates.

Credit Path:
Your design should control and minimize pollutants from entering the building and prevent cross contamination between different areas:

1) After construction and before occupancy, for mechanically ventilated buildings, install <u>filtration media</u> to process both outside air and return air that will be used as supply air. The filtration media shall be <u>13 or better in MERV (Minimum Efficiency Reporting Value)</u>. The filtration system needs to meet <u>ANSI/ASHRAE 52.2, 1999</u> standards.

AND

2) In areas that may have chemical pollutants and possibly dangerous gases or particulates, such as copying/printing room, laundry/ housekeeping areas, garages, prep rooms, science laboratories, art rooms, shops of any kinds, etc., <u>exhaust the areas</u> sufficiently to create a <u>negative pressure</u> when the doors to these areas are closed. For these kinds of areas or spaces, provide <u>hard lid ceilings, deck-to-deck partitions and self-closing doors</u>. Your exhaust rate shall be <u>0.50 cfm/s.f.</u> minimum, and you should not allow air re-circulation. The pressure difference between these spaces and adjacent spaces shall be at least <u>1 Pa</u> (0.004 inches of water) and <u>5 Pa</u> (0.02 inches of water) on average with the doors to these spaces close.

AND

3) Provide <u>permanent entryway systems</u> at all entries regularly used by building users and directly connected to the outdoors; these systems shall be <u>ten feet</u> long minimum in the main travel direction. They are used to capture particulates and dirt, and prevent them from entering the building. You can use <u>grilles, grates, or slotted systems</u> that you can clean from below as entryway systems. You can use <u>roll-out mats</u> only if you have contracted a service organization (or school maintenance staff for schools) to maintain them every week. CS projects with no entryway system cannot obtain this credit.

Submittals:
1) A <u>list of</u> entryway systems
2) A list of areas or rooms to be <u>separated</u>
3) A <u>building maintenance plan</u> including instructions of maintaining and cleaning walk-off mats and entryway systems
4) <u>Details</u> for hard-lid condition and deck-to-deck partition for rooms with contaminants
5) <u>Visual documentation</u> of the size and location of all walk-off mats and permanent entryway systems
6) Review <u>negative pressure calculations</u> at hazardous chemical areas to guarantee appropriate depressurization
7) Keep <u>cutsheets</u> for <u>MERV 13</u> or higher filters

Synergies:
You can use filtration media to remove contaminants from the air in the construction and operation stages of projects. See credits below for high-efficiency filtration criteria:
- IEQc3.1: Construction IAQ Management Plan: During Construction
- IEQc3.2: Construction IAQ Management Plan: Before Occupancy

You can use fan systems to exhaust spaces where chemicals are used. These systems use additional energy and require commissioning. See related prerequisites and credits below:
- EAc1: Optimize Energy Performance
- EAp1: Fundamental Commissioning of the Building Energy Systems
- EAp2: <u>Minimum</u> Energy Performance
- EAc3: Enhanced Commissioning

Ventilation systems should be able to accommodate the filtration media. See related prerequisite and credit below:
- IEQp1: Minimum IAQ Performance
- IEQc1: Outdoor Air Delivery Monitoring

Possible Strategies and Technologies:
1) Provide a separate exhaust system for maintenance areas and facility cleaning areas to control contaminants.
2) Use permanent entry way systems to capture particulates and dirt, and prevent them from entering the buildings.
3) Install high-level filtration media to process both outside air and return air that will be used as supply air. Double check your air handling units to make sure they can accommodate the required pressure drops and filter sizes.

Extra Credit (Exemplary Performance): None

Project Phase: Design Development

<u>LEED Submittal</u> Phase: <u>Design</u>

Related Code or Standard: <u>ANSI/ASHRAE 52.2, 1999</u>

Responsible Party: <u>Contractor, MEP Engineer, and LEED AP</u>
IEQc6.1: Controllability of Systems: Lighting (<u>1</u> Point for NC and Schools. <u>Not Applicable</u> to CS)

Purpose:
Research shows that people like to have control of the lighting system. The purpose of this credit is to give individuals or groups in multi-occupant spaces (like <u>conference areas</u> and <u>classrooms</u>) a high level of controllability of the lighting system for their own preferences, well-being and comfort.

Credit Path for NC:
1) Give controllability of the lighting system to <u>90%</u> or more of the building FTE/users for their preferences and needs.

   AND

   Give controllability of lighting system for <u>ALL shared multi-occupant spaces</u> to meet group preferences and needs.

Credit Path for Schools:
1) For <u>regularly occupied spaces</u> like administrative offices:
   Give controllability of the lighting system to <u>90%</u> or more of the building FTE/users for their preferences and needs.

   AND

   Give controllability of the lighting system for <u>ALL learning spaces</u> like classrooms, art rooms, chemistry laboratories, music rooms, shops, dance and exercise studios, and gymnasiums to meet group preferences and needs.

   OR
2) The lighting system for each <u>classroom</u> needs to have two modes: A/V and general illumination.

   In the **A/V (Audio/Video) mode**, light levels on the walls and ceiling are reduced while a proper light level of about <u>30 footcandles</u> (fc) is maintained on the student desks.

Submittals:
1) A floor <u>plan</u> showing the zoning, location, type of lighting controls, individual and shared work areas, and furniture layout
2) Keep <u>design information</u> on lighting controls, sensors, and task lighting

Synergies:
   See related prerequisite and credit below:
   * EAp1: <u>Fundamental</u> Commissioning of the Building Energy Systems
   * EAp2: <u>Minimum</u> Energy Performance
   * EAc1: Optimize Energy Performance
   * EAc3: <u>Enhanced</u> Commissioning
   * IEQc6.2: Controllability of Systems: Thermal Comfort Systems
   * IEQc8: Daylight and Views

Possible Strategies and Technologies:
1) Use task lighting and lighting controls to provide flexibility for building users.
2) Incorporate controllability into the overall lighting design.
3) Use task lighting and ambient lighting to meet the needs of building users, and still consider and control the building energy use.

Extra Credit (Exemplary Performance): None

Project Phase: Design Development

LEED Submittal Phase: Design

Related Code or Standard: None

Responsible Party: MEP Engineer

IEQc6.2 for NC and Schools, IEQc6 for CS: Controllability of Systems: Thermal Comfort (1 Point for NC, Schools, and CS)

Purpose:
Research shows that people like to have control of the thermal comfort system. The purpose of this credit is to give individuals or groups in multi-occupant spaces (like conference areas and classrooms) a high level of controllability of the thermal comfort system for their own preferences, well-being and comfort.

Credit Path for NC, Schools and CS:
1) Give controllability of the thermal comfort system (for workspaces only in schools) to 50% of the building users for their preferences and needs. Use operable windows instead of comfort control if the occupants' areas are within 10 feet of either side of, and 20 feet inside of, the operable part of a window. Your building's areas of operable window shall meet ASHRAE 62.1-2007 paragraph 5.1, Natural Ventilation (with errata but without addenda), and operable window area shall be 4% or more of the total floor area. Each operable window shall have an individual control system.

AND

2) Give controllability of the thermal comfort system for ALL shared multi-occupant spaces to meet group preferences and needs.

Include **primary factors of comfort** (air speed and air temperature, radiant temperature and humidity) in your consideration for thermal comfort per ASHRAE Standard 55-2004. Your comfort system control shall operate at least one of the primary factors of the building user's local area to be qualified for this credit.

Note:
Non-occupied areas, like <u>closets, equipment rooms, janitor's rooms and storage rooms</u> do <u>NOT</u> need to be included in the calculation for the controllability of thermal comfort.

Additional Requirements for CS:
The CS project team must <u>purchase and/or install</u> the operable windows or mechanical system (or combination of both) to qualify for this credit.

Submittals:
1) For controls at **individual workstations**: a <u>list</u> of the total quantity of thermal controls and individual workstations
2) For controls in **shared spaces**: a <u>description</u> of the thermal controls and a list of shared spaces
3) For CS projects, use the default occupancy count and document the <u>expected tenant number</u> for each floor, and a description of the <u>systems</u> serving each floor

Synergies:
See related prerequisite and credit below:
- EAp1: <u>Fundamental</u> Commissioning of the Building Energy Systems
- EAp2: <u>Minimum</u> Energy Performance
- EAc1: Optimize Energy Performance
- EAc3: <u>Enhanced</u> Commissioning
- EAc5: Measurement and Verification

- IEQc5: Indoor Chemical and Pollutant Source Control Products
- IEQc6.1: Controllability of Systems: Lighting
- IEQc8: Daylight and Views

Possible Strategies and Technologies:
1) Incorporate controllability into the overall thermal comfort design to allow groups in shared spaces or individuals to adjust the system for their preferences and needs.
2) See ASHRAE Standard 55-2004 for **primary factors of comfort** and a process for creating comfort criteria.
3) Develop control strategies to include mechanical systems alone, or operable window or hybrid systems, including both mechanical systems and operable windows.
4) Use local diffusers at overhead levels, desk or floor, individual thermostat controls or individual radiant panel controls.
5) Pay attention to the close relationship between acceptable **<u>indoor air quality</u>** (per ASHRAE Standard <u>62.1</u>-2007) and **thermal comfort** (per ASHRAE Standard <u>55</u>-2004).

Extra Credit (Exemplary Performance): None

Project Phase: Design Development

<u>LEED Submittal Phase:</u> <u>Design</u>

Related Code or Standard:
1) <u>ASHRAE 62.1-2007 (with errata but without addenda)</u>

2) ASHRAE Standard 55-2004

Responsible Party: MEP Engineer

IEQc7.1 for NC and Schools, IEQc7 for CS: Thermal Comfort: Design (1 Point for NC, Schools, and CS)

Purpose:
To protect the well-being of building users and to increase their productivity via a comfortable thermal environment.

Credit Path for NC, Schools, and CS:
Design the building envelope and HVAC systems per ASHRAE Standard 55-2004, Thermal Comfort Conditions for Human Occupancy (with errata but without addenda). You need to show design compliance per Section 6.1.1 Documentation. Use Predicted Mean Vote (PMV) model.

Additional Requirements for Schools:
Show compliance with the "Typical Natatorium Design Condition" per Chapter 4 (place of Assembly) of the ASHRAE HVAC Applications Handbook, 2003 edition (with errata but without addenda) for natatoriums.

Note: A **natatorium** is a structurally separate building containing a swimming pool.

Additional Requirements for CS:
CS projects must allow tenant build-out to meet the criteria of this credit. The CS project team must purchase and/or install the operable windows or mechanical system (or combination of both) to qualify for this credit.

Submittals:
1) Documentation of OPR, including assumptions for occupant clothing and activity level, and comfort criteria
2) Building systems operational procedure and summary, including general information, changeover schedules, recommended seasonal set points, maintenance and operation instructions, inspection and maintenance schedules, building controls, other environmental control systems, etc.
3) Documentation of mechanical engineer's BOD, including HVAC load calculations, diversity consideration, and design assumption
4) Keep design plans, lists and other documentation of all terminal units and registers, including radiant value, flow, type, and any factors that have significant impact on thermal comfort, all occupant-adjustable control locations, and notation of spaces outside comfort-controlled areas.

Synergies:
Thermal comfort is affected by personal factors (clothing and metabolic rate), environment (air speed, air temperature, radiant temperature, and relative humidity). You can control thermal comfort through

passive (natural ventilation) and/or active (HVAC) systems. A combination of both systems can achieve optimum comfort levels and save energy.

See related prerequisite and credit below:
- EAp2: <u>Minimum</u> Energy Performance
- EAc1: Optimize Energy Performance
- EAc5: Measurement and Verification

For commissioning of the thermal comfort features, see related prerequisite and credit below:
- EAp1: <u>Fundamental</u> Commissioning of the Building Energy Systems
- EAc3: <u>Enhanced</u> Commissioning

The following prerequisite and credit also relate to occupants' comfort:
- IEQp1: Minimum IAQ Performance
- IEQc2: Increased Ventilation

Possible Strategies and Technologies:
Set up comfort criteria per ASHRAE Standard 55-2004. Evaluate the primary factors of comfort (air speed and air temperature, radiant temperature and humidity) and relate them to <u>IEQp1, IEQc1 and IEQc2</u>.

Extra Credit (Exemplary Performance): None

Project Phase: Design Development

<u>LEED Submittal</u> Phase: <u>Design</u>

Related Code or Standard: <u>ANSI/ASHRAE Standard 55-2004</u>

Responsible Party: <u>MEP Engineer</u>

IEQc7.2: Thermal Comfort: Verification (<u>1</u> Point for NC and Schools. <u>Not Applicable</u> to CS)

Purpose:
To follow up and evaluate the building's thermal comfort.

Credit Path for NC and Schools:
Conduct a thermal comfort survey of building users (students of grade 6 and above as well as adults) within <u>6 to 18 months</u> after occupancy. The survey should be anonymous, and it should identify the <u>problems</u> related to thermal comfort and <u>overall satisfaction</u> with the thermal system performance. If <u>less than 80%</u> of the building occupants are satisfied, then develop a plan to correct the problems. Your corrective actions should include measuring the environmental variables in the problem areas per <u>ASHRAE Standard 55-2004</u> (with errata but without addenda).

Additional Requirements for NC:
Install a permanent monitoring system to make sure the building performs as designed and meets the comfort criteria per IEQc7.1.

Residential projects do not qualify for this credit.

Submittals:
1) Create and administer <u>a thermal comfort survey for building occupants</u>
2) If <u>20%</u> or more of the building occupants are not satisfied with the thermal comfort, <u>write a plan for corrective action</u>

Synergies:
Thermal comfort is affected by personal factors (clothing and metabolic rate), environment (air speed, air temperature, radiant temperature, and relative humidity). You can control thermal comfort through passive (natural ventilation) and/or active (HVAC) systems. A combination of both systems can achieve optimum comfort levels and save energy.

See related prerequisite and credit below:
- EAp1: <u>Fundamental</u> Commissioning of the Building Energy Systems
- EAc3: <u>Enhanced</u> Commissioning
- EAc5: Measurement and Verification

The following prerequisite and credit also relate to occupants' comfort:
- IEQp1: Minimum IAQ Performance
- IEQc2: Increased Ventilation
- IEQc6.2: Controllability of Systems: Thermal Comfort
- IEQc7.1: Thermal Comfort: Design

Possible Strategies and Technologies:
Use the principles in ASHRAE Standard 55-2004 to develop criteria for thermal comfort, and a plan for monitoring and correcting performance of the thermal comfort systems.

Extra Credit (Exemplary Performance): None

Project Phase: Design Development

<u>LEED Submittal</u> Phase: <u>Design</u>

Related Code or Standard: <u>ANSI/ASHRAE Standard 55-2004</u>

Responsible Party: <u>MEP Engineer</u>

IEQc8.1: Daylight and Views: Daylight (<u>1</u> Point for NC and CS, <u>1-3</u> Points for Schools)

Purpose:

To connect the inside and outside of the building by introducing views and daylight to the regularly occupied areas inside the building.

Credit Paths for Schools:

% of Daylight in Classrooms	Points
75%	1
90%	2

You can obtain <u>1</u> extra point for providing daylight for <u>75%</u> of all <u>other</u> regularly occupied spaces, but you need to achieve at least 1 point for classroom spaces to get the extra point.

Credit Paths for NC and CS:

% of Daylight in Regularly Occupied Spaces	Points
75%	1

Credit Paths for NC, Schools and CS:

1) **Prescriptive**

   For side-lighting daylight area:

   0.15 < VLT × WFR <0.18

   VLT is visual light transmittal
   WFR is window-to-floor area ratio

- The window area must be a minimum of <u>30 inches</u> above the floor to be included in the calculation.
- The <u>horizontal</u> distance from the <u>edge</u> of the daylight area to the <u>bottom</u> of the wall (where the window is located) shall be <u>less than two times</u> the <u>vertical</u> height from the <u>bottom</u> of the wall to the top of window <u>mullion</u>.
- If you draw a line from the <u>edge</u> of the side-lighting daylight area to the <u>top</u> of window <u>glass</u>, the ceiling shall be above this line so as not to block daylight.

   For top-lighting daylight area:
- The top-lighting daylight area under the skylight is the area directly under the <u>outline</u> of the skylight <u>PLUS</u> the Lesser of:

   <u>70%</u> of the ceiling height

   OR
   <u>50%</u> of the distance to the closest of the adjacent skylights

   OR

The distance to any permanent opaque partition (show VLT for transparent partitions) that is at a distance of 70% or more of the gap between the ceiling and the top of the partition.

- Skylight coverage shall be 3% to 6% of the roof area with a VLT of at least 0.5
- The distance between skylights shall be 1.4 times the ceiling height or less.
- If you use a skylight diffuser, it shall have a measured haze value of 90% or more when tested per ASTM D1003. In your design, you should avoid direct sightlines to the skylight diffuser.

OR
2) **Computer Simulation**

Use a computer simulation to show that applicable spaces have at least 25 horizontal footcandles (fc) and do not exceed 500 fc of daylight illumination. This simulation should be based on a clear sky situation, on September 21, at 9 a.m. and noon, at 30 inches above the finish floor. If your design incorporates view-preserving automated shades for glare control, you just need to meet the 25 fc minimum requirement.

OR
3) **Measurement**

Use records of measurements of indoor lights to show that applicable spaces have at least 25 horizontal footcandles of daylight illumination. Take the measurements on a 10-foot grid and mark the measurements on the floor plans.

For all three options, only apply the s.f. of the portions of the spaces with at least 25 horizontal footcandles of daylight illumination towards the 75% required areas.

For all three options, use control devices or redirect daylight to avoid high-contrast conditions and prevent impeding visual tasks. Exceptions to this rule will be considered based on their merits.

4) **Combination**

Any combination of the three paths above can be used. If the combination results in 75% of the regularly occupied areas meeting the requirements, your project can get one point. If the combination results in 90% of the regularly occupied areas meeting the requirements, your project can get two points.

In all options, only the area related to the spaces or rooms meeting the requirements can be included in the calculation.

In all options, you can use glare control devices to maximize daylight for interior spaces while avoiding high-contrast conditions.

Note:
1) Include all regularly occupied spaces like cafeterias, meeting rooms and office spaces, but do NOT include spaces like closets, utility rooms, bathrooms, copy rooms, laundry rooms, restrooms, equipment rooms, janitor's rooms, or storage rooms in the area calculations for daylight and views.
2) Include window areas between 2'-6" AFF and 7'-6" AFF, and glazing areas above 7'-6" AFF. Do NOT include areas for any openings below 2'-6" AFF.

Submittals:
1) Create floor plans, elevations, sections, etc. to show the glare control methods used
2) Keep floor plans, elevations, sections, etc. to show a qualifying amount of daylight for regularly occupied spaces
3) A spreadsheet to document daylight factors outlined in calculations, and changes in design
4) Update the computer model continuously if using the simulation method

Synergies:
Increasing window size to gain more daylight can also increase views. See related credit below:
- IEQc8.2: Daylight and Views

Bigger windows have a direct impact on building energy performance. Daylighting controls can maximize energy savings.
See related prerequisite and credit below:
- EAp2: Minimum Energy Performance
- EAc1: Optimize Energy Performance
- IEQc6: Controllability of Systems

Possible Strategies and Technologies:
1) Use interior and exterior permanent shading devices, adjust building orientation, increase building perimeter, design shallow floor plates, properly design the size, number and locations of building openings, consider neighboring building and trees, use courtyards, atriums, clerestory windows, north-facing skylight, interior light (reflecting) shelf, and use automatic photocell-based control and high performance glazing to maximize daylight for interior spaces while avoiding high-contrast conditions.
2) Use light fixtures with sensors and dimming controls.
3) Use a computer model, physical model or manual calculation to evaluate daylight footcandle levels.

Extra Credit (Exemplary Performance):
For NC and CS projects, you can get one extra innovation point by providing daylight for a minimum of 95% of the regularly occupied areas per the credit requirements and guidelines.

For Schools:
1) You can obtain 1 extra point for providing daylight for 75% of all other regularly occupied spaces, but you need to achieve at least 1 point for classroom spaces to get this extra point for the other spaces.
2) You may obtain 1 extra point for providing daylight for 90% of all classrooms and 95% of all other regularly occupied spaces.

Project Phase: Schematic Design

LEED Submittal Phase: Design

Related Code or Standard:
ASTM D1003-07e1, Standard Testing Method for Haze and Luminous Transmittance of Transparent Plastics

Responsible Party: Lighting Designer, Architect and LEED AP

IEQc8.2: Daylight and Views: Views (1 Point for NC, Schools, and CS)

Purpose:

To connect the inside and outside of the building by introducing views and daylight to regularly occupied areas inside the building.

Credit Path for NC, Schools, and CS:

Provide views (direct line of sight to the outside) for 90% of the regularly occupied spaces. The line of sight should be via vision glazing at a height between 2'-6" and 7'-6" AFF. An area needs to meet the two criteria below to be included as an area with a view:

1) In plan view, be able to draw a line of sight from the area to the perimeter vision glazing (between 2'-6" and 7'-6" AFF).

2) In section view, draw a line of sight from the area to the perimeter vision glazing. Use 42" as the average seated eye height.

You can draw a line of sight through interior glazing. If at least 75% of a private office has a view, then you can count the entire area of the office as an area with a view. If a space has multiple occupants, then you can only count the actual areas with a view.

Additional Requirements for CS:

CS projects shall include a feasible tenant layout in the analysis for this credit. The layout shall be per the default occupant counts or similar data.

Submittals:

1) Keep floor plans, elevations, sections, etc. to show locations of regularly occupied spaces with views

2) A spreadsheet to document view area outlined in calculations, and any changes in design

Synergies:

Same as IEQ8.1, except replacing IEQ8.1 with IEQ8.2.

Possible Strategies and Technologies:

1) Design the space to maximize view and daylight.

2) Use interior glazing, interior shading device, low partition heights and automatic photo-cell controls.

Extra Credit (Exemplary Performance):

If you achieve two out of the four items below, you can obtain an extra ID point:

1) At least 90% of the regularly occupied spaces have access to unobstructed views within a distance of three times the vision window head height.

2) At least 90% of the regularly occupied spaces have more than one sight line to vision glazing in different directions at 90 degrees or more apart.

3) At least 90% of the regularly occupied spaces have views to two of the three choices: human activities; vegetation; objects 70 feet or more from the exterior of the glazing.

4) At least 90% of the regularly occupied spaces have access to views with a view factor of 3 or more.

Project Phase: Schematic Design

LEED Submittal Phase: Design

Related Code or Standard: None

Responsible Party: Lighting Designer, Architect and LEED AP

IEQc9: Enhanced Acoustical Performance (1 Point for Schools Only)

Note: Detailed discussions have been omitted since this credit is for schools only, and is unlikely to be tested on the LEED Green Associate Exam.

IEQc10: Mold Prevention (1 Point for Schools Only)

Note: Detailed discussions have been omitted since this credit is for schools only, and is unlikely to be tested on the LEED Green Associate Exam.

IEQ Summary and Mnemonics:

IEQ Credit Name	Extra Credit	Responsible Party
IEQp1: **M**inimum IAQ Performance (Required for NC, Schools and CS)	0	MEP Engineer
IEQp2: **E**nvironmental Tobacco Smoke (ETS) Control (Required for NC, Schools and CS)	0	LEED AP, Owner, and MEP Engineer
*IEQp3: Minimum Acoustical Performance (Required for Schools only, N/A for NC & CS)	0	Mechanical Engineer and Architect
IEQc1: **O**utdoor Air Delivery Monitoring (1 Point for NC, Schools, and CS)	0	MEP Engineer
IEQc2: **I**ncreased Ventilation (1 Point for NC, Schools, and CS)	0	Contractor and MEP Engineer
*IEQc3.1 for NC and Schools, and *IEQc3 for CS: **C**onstruction IAQ Management Plan: During Construction (1 Point for NC, Schools, and CS)	0	Contractor and MEP Engineer

*IEQc3.2: Construction IAQ Management Plan: Before Occupancy (1 Point for NC and Schools, N/A for CS)	0	Contractor and MEP Engineer
IEQc4.1: Low-Emitting Materials: Adhesive and Sealants (1 Point for NC, Schools and CS)	0	Contractor and Architect
IEQc4.2: Low-Emitting Materials: Paints and Coatings (1 Point for NC, Schools, and CS)	0	Contractor and Architect
IEQc4.3: Low-Emitting Materials: Carpet Systems (1 Point for NC, Schools, and CS)	0	Contractor and Architect
IEQc4.4: Low-Emitting Materials: Composite Wood and Agrifiber Products (1 Point for NC, Schools, and CS)	0	Contractor and Architect
*IEQc4.5 Low-Emitting Materials: Furniture and Furnishings Systems (1 Point for Schools Only, N/A for NC & CS)	0	Contractor and Architect
*IEQc4.6 Low-Emitting Materials: Ceiling and Wall Systems (1 Point for Schools Only, N/A for NC & CS)	0	Contractor and Architect
IEQc5: Indoor Chemical and Pollutant Source Control Products (1 Point for NC, Schools, and CS)	0	Contractor, MEP Engineer, and LEED AP
*IEQc6.1: Controllability of Systems: Lighting (1 Point for NC and Schools. N/A to CS)	0	MEP Engineer
*IEQc6.2 for NC and Schools, *IEQc6 for CS: Controllability of Systems: Thermal Comfort (1 Point for NC, Schools, and CS)	0	MEP Engineer
*IEQc7.1 for NC and Schools, *IEQc7 for CS: Thermal Comfort: Design (1 Point for NC, Schools, and CS)	0	MEP Engineer
*IEQc7.2: Thermal Comfort: Verification (1 Point for NC and Schools. N/A to CS)	0	MEP Engineer

IEQc8.1: **D**aylight and Views: Daylight (1 Point for NC and CS, 1-3 Points for Schools)	For NC and CS projects: You can get one extra innovation point by providing daylight for a minimum of 95% of the regularly occupied areas per the credit requirements and guidelines. For Schools: 1) You can obtain 1 extra point for providing daylight for 75% of all other regularly occupied spaces, but you need to achieve at least 1 point for classroom spaces to get this extra point for the other spaces. 2) You may obtain 1 extra point for providing daylight for 90% of all classrooms and 95% of all other regularly occupied spaces.	Lighting Designer, Architect, and LEED AP
IEQc8.2: Daylight and Views-Views (1 Point for NC, Schools, and CS)	If you achieve two out of the four items below, you can obtain an extra ID point: 1) At least 90% of the regularly occupied spaces have access to unobstructed views within a distance of three times the vision window head height. 2) At least 90% of the regularly occupied spaces have more than one sight line to vision glazing in different directions at 90 degrees or more apart. 3) At least 90% of the regularly occupied spaces have views to two of the three choices: 4) human activities; vegetation; objects 70 feet or more from the exterior of the glazing. 5) At least 90% of the regularly occupied spaces have access to views with a view factor of 3 or more.	Lighting Designer, Architect, and LEED AP
*IEQc9: Enhanced Acoustical Performance (1 Point for Schools Only)	1 extra ID credit if the project achieves 1) An indoor noise level of 35 dBA OR 2) A noise level of 55 dBA for playground background, and a 65 dBA noise level for athletic fields and all other school grounds.	MEP Engineer and Architect
*IEQc10: Mold Prevention (1 Point for Schools Only)	Case-by-case	MEP Engineer

Mnemonics:

ME

Oh! Ian Catches LIST D (See bold and underlined letters in credit names on table above also). Of course, you can create your own **mnemonics** to help you memorize them.

Note: * indicates prerequisite or credit NOT applicable to all LEED rating systems. Detailed discussions have been omitted since the information is unlikely to be tested on the LEED Green Associate Exam.

F. Innovation and Design Process (ID)

Because this credit category is relatively simple, we omit the overall discussion.

IDc1: Innovation in Design (1-5 points for NC and CS. 1-4 points for Schools)

Purpose:
To reward design teams and projects for innovative performance in green building categories NOT covered by LEED and/or exceptional performance above and beyond the LEED requirements.

Credit Paths for NC, Schools, and CS:
You can obtain credits through any combinations of the paths below:

1) **Innovation in Design (1-5 points for NC and CS. 1-4 points for Schools)**
   In writing, provide the intent, requirements, proposed submittals and design approach for the proposed innovation credit. You need to show:
   a. Innovative performance in green building categories NOT covered by an existing LEED credit.
   b. The GBCI will award one point for each innovation achieved.
   c. The **criteria** are: quantitative performance improvements for environmental benefit; comprehensive specification or process (applicable to the entire project, not just a portion); substantially better than typical sustainable practice and applicable to other projects.

OR

2) **Exemplary Performance (1-3 points)**
   a. Exceptional performance above and beyond the LEED-NC requirements for an existing credit. For example, in a number of LEED credit categories, you can get one innovation point by doubling the requirements or reaching the next threshold.
   b. The GBCI will award one point for each exemplary performance achieved.
   c. A maximum of three points can be achieved through this path.

   Note: The benefits from different paths need to be additive.

3) **Pilot Credit (1 point)**
   a. Go to the USGBC Pilot Credit Library:

      http://www.usgbc.org/PilotCreditLibrary

   b. Register as a pilot credit participant and complete the required documents.

   c. The project team can seek more than 1 pilot credits, but the USGBC will ONLY award 1 pilot credit for each LEED project.

Submittals:
1) Documentation for the <u>process</u> of the project team to achieve environmental benefits or exceptional performance above and beyond the LEED-NC requirements in an existing credit, or innovative performance in green building categories <u>NOT</u> covered by an existing LEED credit
2) <u>Track implementation and development</u> of the innovative and exceptional strategies used
3) For **CS** projects, indicate the <u>scope of project</u> that the ID point <u>covers</u>

Synergies:
See extra credit (exemplary performance) section of each credit.

Possible Strategies and Technologies:
1) Try to <u>substantially</u> exceed the LEED performance credit (like WE or IEQ).
2) Use measures that can show a <u>quantifiable</u> health and/or environmental benefits, and a <u>comprehensive</u> approach.
4) Try to select innovative performance that is applicable for <u>other</u> projects.
5) If an innovation point is awarded, the GBCI <u>does not guarantee</u> it will be awarded for future projects.
6) You need to do one <u>separate</u> submittal and narrative statement for each point.

Extra Credit (Exemplary Performance):
None (IDc1 IS the extra ID credit for other LEED credit categories).

Project Phase: Varies

<u>**LEED Submittal Phase:**</u> <u>Construction or Design</u>

Related Code or Standard: None

Responsible Party: Varies (<u>All disciplines, Owner, and Contractor</u>)

IDc2: LEED Accredited Professional (1 point for NC, Schools, and CS)

Purpose:
To encourage and support the design integration dictated by LEED and to streamline the certification and application process.

Credit Path for NC, Schools and CS:
To get this credit, at least one <u>principal</u> project team member (not a <u>junior</u> team member) shall be a LEED AP (Accredited Professional).

Submittals:
Confirmation of the LEED AP status of at least one principal of the team.

Synergies: None

Possible Strategies and Technologies:
1) Try to assign the LEED AP as the facilitator for the integrated design and construction process.
2) Educate your team early about the LEED rating system and green buildings.

Extra Credit (Exemplary Performance): None

Project Phase: ALL phases

LEED Submittal Phase: Construction

Related Code or Standard: LEED AP. See GBCI.org

Responsible Party: LEED AP, Owner, and Contractor

IDc3: The School as a Teaching Tool (1 point for Schools Only)

Note: Detailed discussions have been omitted since this credit is for schools only, and is unlikely to be tested on the LEED Green Associate Exam.

ID Summary and Mnemonics:

ID Credit Name	Extra Credit	Responsible Party
IDc1: Innovation in Design (1-5 points for NC and CS. 1-4 points for Schools)	None (ID credits ARE the extra ID credits for other LEED credit categories).	Varies (All disciplines, Owner, and Contractor)
IDc2: LEED Accredited Professional (1 point for NC, Schools, and CS)	Same as above	LEED AP, Owner, and Contractor
*IDc3: The School as a Teaching Tool (1 point for Schools only, N/A for NC & CS)	Same as above	Teachers and the project team

Mnemonics:
I Love Sandy. (See bold and underlined letters in credit names on table above also). Of course, you can create your own **mnemonics** to help you memorize them.

Note: * indicates prerequisite or credit NOT applicable to all LEED rating systems. Detailed discussions have been omitted since the information is unlikely to be tested on the LEED Green Associate Exam.

G. Regional Priority (RP)

Because this credit category is relatively simple, we omit the overall discussion.

RPc1: Regional Priority (1-4 points for NC, Schools and CS)

Purpose:
To encourage the achievement of credit for addressing regional environmental priorities.

Credit Paths for NC, Schools and CS:
See the RP credit database at
http://www.usgbc.org

You can obtain one to four credits out of the six potential RP points. You can choose which four credits you want to pursue for your project.

The USGBC has prioritized the projects located in the US, the US Virgin Islands, Guam, and Puerto Rico. Project teams for other international projects can check the database above for eligible RP points.

Submittals:
See RP database at www.usgbc.org and LEED Online

Synergies:
See RP database at www.usgbc.org

Possible Strategies and Technologies:
1) The available RP credits will show up at LEED Online after you enter your project's specific information like zip code.
2) Coordinate with other credits and choose which four RP credits you want to pursue for your project.

Extra Credit (Exemplary Performance): None

Project Phase: ALL phases

LEED Submittal Phase: Construction

Related Code or Standard: None

Responsible Party: LEED AP

Chapter 5

LEED Green Associate Exam
(LEED-GA) Sample Questions,
Mock Exam, Answers and
Exam Registration

I. LEED Green Associate Exam sample questions

Use these sample questions to prepare for the mock exam and the real exam. They will give you an idea what the USGBC is looking for on the LEED Green Associate Exam, and how the questions will be asked. These sample questions are quite easy. If you can answer 80% of the sample questions correctly, you are ready to take the mock exam in this chapter. The 80% passing score is based on feedback from previous readers. You need to read the study material the in previous chapters one or more times to become familiar with it and then take the mock exam. You really need to read this book several times and <u>MEMORIZE</u> the important information before you do the sample questions and mock exam. Just like on the real exam, sometimes a question may ask you to pick two or three correct answers out of four, or four correct answers out of five (some LEED exam questions have five choices). This means that if you do not know any one of the correct answers, you will probably get the overall answer wrong. You need to know the LEED system very well to get the correct answer.

One <u>exception</u> that I make and it may be different from the real LEED Green Associate Exam: I still break the main LEED credit into separate credits, like SSc1, MRc3, etc. This is because I believe most people will take the LEED AP+ or LEED AP with specialty exam at some point after they take the LEED Green Associate Exam.

By organizing this book by each specific LEED credit, I make it <u>easier for you to support a LEED project team</u> because people will talk about SSc8, MRc6, etc. when they do real LEED projects.

Another benefit is that you just need to spend <u>50%</u> more effort to prepare for Part II of the LEED AP+ exam since you already know most of specific LEED credits: the basic and fundamental credit system is very similar for ALL LEED systems.

If I did not break the main LEED category into separate credits, you will have to <u>double</u> your effort to prepare for Part II of the LEED AP+ exam since you will have to <u>MEMORIZE all information twice</u>: once for each main LEED category and then memorize it again per each specific LEED credit.

I tried to organize the LEED information by each main LEED category only when I first started to write this book, and I ended up with 80 to 90 items for each main LEED category. It was very confusing. So I decided to organize the information by specific LEED credit.

It is much easier to <u>understand, digest and memorize</u> the information by <u>specific</u> LEED credit instead of just by the main LEED category.

1. With regard to the Optimize Energy Performance credit, which of the following statements is correct?
   a. Compare your building performance to the baseline building performance.
   b. Compare baseline building performance to ASHRAE Standard 90.1-2007 (with errata but without addenda).
   c. Compare baseline building performance to ASHRAE Standard 90.1-2004.
   d. Compare baseline building performance to ASHRAE Standard 90.1-2003.

2. Which of the following factors does not improve human comfort?
   a. Air temperature.
   b. Ventilation.
   c. Radiation exchange.
   d. All of the above.

3. The standard used for Measurement and Verification is:
   a. ASHRAE Standard 90.1-2007 (with errata but without addenda).
   b. The Department of Energy Verification Protocol.
   c. A signed statement from the designer.
   d. International Performance Measurement and Verification Protocol.

4. Optimal IAQ performance can:
   a. Create savings in electrical bills.
   b. Improve the productivity and health of building occupants.
   c. Cause higher operation cost.
   d. Create higher rents.

5. What is the best way to control Environmental Tobacco Smoke (ETS)?
   a. Ban smoking inside the building.
   b. Place exterior smoking area at least 25 feet away from operable windows, entrances and air intakes.
   c. All interior spaces shall have negative pressure.
   d. Both a and b.

6. With regard to the Construction IAQ Management Plan, if your project uses permanently installed air handlers during construction, you should use
   a. Filtration media with a minimum Efficiency Reporting Value (MERV) of 6 at each return air grille.
   b. Filtration media with a minimum Efficiency Reporting Value (MERV) of 7 at each return air grille.
   c. Filtration media with a minimum Efficiency Reporting Value (MERV) of 8 at each return air grille.
   d. Filtration media with a minimum Efficiency Reporting Value (MERV) of 9 at each return air grille.

7. For Indoor Chemical and Pollutant Source Control, you need to
   a. Physically isolate activities related to chemical use.
   b. Submit drawings and cutsheets for the plumbing systems in chemical mixing area.
   c. Design the exterior sidewalk and pavement to drain away from the building at 2% minimum slope.
   d. None of the above.

8. Placing a lighting control in a hallway does not help you with Controllability of Systems because
   a. You need to make sure the hallway does not have a dead end corridor that is over 20 feet in length.
   b. You also need to add temperature control.
   c. You also need to make sure the hallway has a view to the outside.
   d. None of the above.

9. Which of the following standard(s) is mentioned with regard to Increased Ventilation?
   a. ASHRAE Standard 52.2-1999.
   b. ASHRAE Standard 62.1-2007.
   c. ASHRAE Standard 90.1-2007 (with errata but without addenda).
   d. SCAQMD Rule 1168.

10. Which of the following standard(s) is (are) mentioned in Light Pollution Reduction?
    a. ASHRAE/IESNA Standard 90.1-2007 (with errata but without addenda).
    b. IESNA RP-33.
    c. International Dark Sky Association Lighting Standard.
    d. Both a and b.

11. California Title 24-2005 is considered to be equal to ASHRAE/IESNA Standard 90.1-2007 (with errata but without addenda) for the following LEED-NC credit(s):
    a. Minimum Energy Performance – EAp2.
    b. Optimize Energy Performance – EAc1.
    c. Green Power-EAc6.
    d. All of the above.

12. Which of the following is not a rapidly renewable material?
    a. Bamboo flooring.
    b. Linoleum Flooring
    c. Glass Windows
    d. Wool Carpeting

13. What does "Xeriscape" mean?
    a. Drip Irrigation to save water.
    b. "Dry Landscape" design by using plants that use little or no water.
    c. Recycle existing plants on the project site.
    d. Reuse graywater for landscape irrigation.

14. Which of the following is a responsibility for the contractor to support the LEED documentation process?

    a. Document and provide calculations for waste diverted from landfill.

    b. Maintain a submittal log.

    c. Maintain a RFI log.

    d. Provide written documentation to justify a change order for rough grading.

15. Rapidly renewable material may also qualify for which credit(s)?

    a. MRc3, Material Reuse

    b. MRc5, Regional Materials

    c. MRc7, Certified Wood.

    d. All of the above.

16. Which of the following three statements are correct? (Choose 3)

    a. For non-residential projects, water closet uses per day per female FTE is 3.

    b. For residential projects, water closet uses per day per female is 5.

    c. The Flow Rate for a conventional water closet is 1.8 gpf.

    d. The Flow Rate for a low-flow water closet is 1.1 gpf.

17. Which of the following is the standard for Low-Emitting Sealants, Sealant Primers and Adhesives?

    a. SCAQMD (South Coast Air Quality Management District) Rule #1168, VOC limits effective 7/1/2005.

    b. Green Seal Standard for Commercial Adhesives GS-36 requirements with an effective date of 10/19/2000

    c. Green Label Plus

    d. Green Guard

18. Why should a developer locate a green building in a previously developed urban area?

    a. It is close to public transportation.

    b. It can use existing community services.

    c. It is close to existing utilities.

    d. a and b

    e. a, b and c

19. Blackwater is water drained from a:

    a. kitchen sink

    b. toilet.

    c. both a and b.

    d. None of the above

20. With regard to ozone depletion potential (ODP), the order from high ODP to low ODP is:

    a. CFC> HFC>HCFC

    b. CFC>HCFC>HFC

    c. HCFC>HFC>CFC

    d. HCFC> CFC>HFC

II. Answers for the LEED Green Associate Exam sample questions

1. Answer: a
2. Answer: c
3. Answer: d
4. Answer: b
5. Answer: d
6. Answer: c
7. Answer: a
8. Answer: d
9. Answer: b
10. Answer: d
11. Answer: d
12. Answer: c
13. Answer: b
14. Answer: a
15. Answer: d
16. Answer: a, b and d
17. Answer: a
18. Answer: e
19. Answer: c
20. Answer: b

III. LEED Green Associate Mock Exam

1. What is the maximum number of points that you can get for LEED-CS?
    a. 50 points
    b. 60 points
    c. 80 points
    d. 110 points

2. What are the different levels of LEED-Schools Certification
    a. LEED-SILVER, LEED-GOLD, LEED-PLATINUM
    b. LEED-CERTIFIED, LEED-SILVER, LEED-GOLD, LEED-PLATINUM
    c. LEED-BRONZE, LEED-SILVER, LEED-GOLD, LEED-PLATINUM
    d. LEED-BRONZE, LEED-SILVER, LEED-GOLD

3. How many categories are there for LEED-NC? What are they?
    a. There are 4. They are Sustainable Site Development (SS), Energy and Atmosphere (EA), Indoor Environmental Quality (IEQ), and Regional Priority (RP).
    b. There are 4. They are Sustainable Site Development (SS), Water Efficiency (WE), Energy and Atmosphere (EA), and Indoor Environmental Quality (IEQ).
    c. There are 6. They are Sustainable Site Development (SS), Water Efficiency (WE), Energy and Atmosphere (EA), Materials and Resource (MR), Indoor Environmental Quality (IEQ), and Regional Priority (RP).
    d. There are 7. They are Sustainable Site Development (SS), Water Efficiency (WE), Energy and Atmosphere (EA), Materials and Resource (MR), Indoor Environmental Quality (IEQ), Innovation and Design Process (ID), and Regional Priority (RP).

4. When was the LEED-NC 2.2 EA section revised?
    a. May 25, 2006
    b. July 26, 2006
    c. June 26, 2006
    d. March 23, 2006

5. What does LEED stand for?
    a. Leadership in Energy and Environmental Design.
    b. Leadership in Efficient Environmental Design.
    c. Leadership in Energy Efficiency Design.
    d. Leadership in Engineering and Environmental Design.

6. When was the USGBC founded?
    a. 1992
    b. 1993
    c. 1996
    d. 1998

7. Which of the following professionals is not involved in green building design?
   a. Architect
   b. Mechanical Engineer
   c. Landscape Architect
   d. Equipment Vendor

8. When was LEED-NC v2.2 published?
   a. 2003
   b. 2002
   c. 2001
   d. 2000

9. For LEED-NC, how many points are needed for Silver Certification?
   a. 30 points
   b. 40 points
   c. 50 points
   d. 60 points

10. For Construction Activity Pollution Prevention, ESC stands for:
    a. Environmental Safety Council
    b. Environment and Sedimentation Control
    c. Engineering Safety Council
    d. Erosion and Sedimentation Control

11. For Development Density and Community Connectivity, you need to construct a building:
    a. On a previously developed site AND within ¼ mile of 10 basic services AND within a neighborhood with 10 units per acre net on average or within a residential zone.
    b. On a previously developed site AND within one mile of 10 basic services AND within a neighborhood with 10 units per acre net on average or within a residential zone.
    c. On a previously developed site AND within ½ mile of 10 basic services AND within a neighborhood with 60 units per acre net on average or within a residential zone.
    d. On a previously developed site AND within ½ mile of 10 basic services AND within a neighborhood with 10 units per acre net on average or within a residential zone.

12. To get one point for Alternative Transportation, your building needs to be:
    a. located within ½ mile of a subway or rail station.
    b. located within ¼ mile of two bus stops.
    c. located within ¼ mile of one or more bus stops for two or more bus lines.
    d. a and c.

13. You can get one point for Alternative Transportation, by (Choose 2):
    a. Providing secure bike storage and/or rack for 3% of the building users.
    b. Providing secure bike storage and/or rack for 5% of the building users.
    c. Providing secure bike storage and/or rack for 5% of the building users within 200 yards of the building entrance.
    d. Provide shower and changing facilities for 0.5% of FTE.

14. You can get three points for EAc1: Optimize Energy Performance by:
    a. Improving a new building's energy performance by 12%.
    b. Improving an existing building's energy performance by 12%.
    c. Improving a new building's energy performance by 14%.
    d. Both a and b.

15. You can get one point for Innovative Wastewater Technology by:
    a. Treating, infiltrating or using 30% of the waste water on-site.
    b. Treating, infiltrating or using 50% of the waste water on-site.
    c. Treating, infiltrating or using 60% of the waste water on-site.
    d. None of above.

16. You can get one point for Regional Material if:
    a. At least 20% (based on cost) of the total materials have been recovered, harvested or extracted within 500 miles of your project site.
    b. At least 10% (based on cost) of the total materials have been recovered, harvested or extracted within 500 miles of your project site.
    c. At least 15% (based on cost) of the total materials have been recovered, harvested or extracted within 500 miles of your project site.
    d. At least 30% (based on cost) of the total materials have been recovered, harvested or extracted within 500 miles of your project site.

17. You can get one point for the EQ Category by:
    a. Providing daylight for 75% of the space.
    b. Providing daylight for 90% of the space.
    c. Providing a view for 90% of the space.
    d. Both a and c.

18. What is the maximum number of points that you can get for the ID credit?
    a. 6 points.
    b. 5 points.
    c. 4 points.
    d. 3 points.

19. If you have only one LEED AP on your team as a job captain, how many point(s) can you get?
    a. 0 points.
    b. 1 point.
    c. 2 points.
    d. 3 points.

20. ASHRAE means
    a. American Society of Heating, Refrigerating, Air Conditioning and Environment.
    b. American Society of Heating, Refrigerating, Air Conditioning and Engineering.
    c. American Society of Heating, Refrigerating and Air Conditioning Engineers.
    d. None of above.

21. Which two of the following are not potable water per LEED?
    a. Reclaimed irrigation water.
    b. Water in detention pond.
    c. Water from municipal water system.
    d. Well water.

22. You are designing a 50,000 s.f. building with five equal floors in an area with no zoning requirements. To get one point for Site Development, you need to provide:
    a. 50,000 s.f. of vegetated open space adjacent to your building.
    b. 25,000 s.f. of vegetated open space adjacent to your building.
    c. 12,500 s.f. of vegetated open space adjacent to your building.
    d. 10,000 s.f. of vegetated open space adjacent to your building.

23. For Fundamental Commissioning of the Building Energy System, you are working on a project with 45,000 gross s.f. and you are looking for a CxA. Which of the following statements is correct?
    a. The CxA shall be independent of the project's design team.
    b. The CxA shall be independent of the project's construction management.
    c. Both a and b.
    d. The CxA can be a qualified person on the construction or design team with the necessary experience.

24. To get three points for On-Site Renewable Energy for a NC project, you must:
    a. provide 3% renewable energy (based on annual building energy cost).
    b. provide 4% renewable energy (based on annual building energy cost).
    c. provide 5% renewable energy (based on annual building energy cost).
    d. provide 8% renewable energy (based on annual building energy cost).

25. For Low-Emitting Materials, VOC means:
    a. Volatile Organic Compounds.
    b. Volatile Organic Components.
    c. Volume of Circulation.
    d. Validity of Credit.

26. For Outdoor Air Delivery Monitoring, the monitoring locations of carbon dioxide concentrations in densely occupied space for mechanically ventilated spaces shall be:
    a. 2'-6" to 6'-6" above the floor.
    b. 3' to 6' above the floor.
    c. 3'-6" to 6'-6" above the floor.
    d. 4'-6" to 6'-6" above the floor.

27. To get one point for MRc1.1: Building Reuse for a school, you should:
    a. Maintain 75% of existing floors, walls and roof.
    b. Maintain 80% of existing floors, walls and roof.
    c. Maintain 85% of existing floors, walls and roof.
    d. Maintain 90% of existing floors, walls and roof.

28. To get one point for MRc1.2, you should:
    a. Maintain 25% of interior non-structural elements.
    b. Maintain 50% of interior non-structural elements.
    c. Maintain 75% of interior non-structural elements.
    d. Maintain 95% of interior non-structural elements.

29. To get two points for Construction Management, you should:
    a. divert 50% from the landfill waste stream in your construction management plan.
    b. divert 75% from the landfill waste stream in your construction management plan.
    c. divert 80% from the landfill waste stream in your construction management plan.
    d. divert 90% from the landfill waste stream in your construction management plan.

30. To get one point for Enhanced Refrigerant Management, you can choose:
    a. Not to use refrigerants.
    b. Choose HVAC&R and refrigerants that have no or reduced emission (below the maximum threshold) of compounds contributing to ozone depletion and global warming.
    c. Use small HVAC units and other cooling equipment containing less than 0.50 lbs. of refrigerant.
    d. All of above and you should not use ozone-depleting substances for your fire suppression systems.

31. What is the minimum renewable energy contract term for Green Power?
    a. One year.
    b. Two years.
    c. Three years.
    d. Four years.

32. If you reduce your building's water use by 30%, how many points can you get from Water Use Reduction?
    a. 2 points.
    b. 1 point.
    c. 3 points.
    d. None of above.

33. If you reduce your landscape irrigation water use by 20%, how many points can you get for WEc3, Water Use Reduction?
    a. 2 points.
    b. 1 point.
    c. 0 points.
    d. None of above.

34. If you reduce your potable water use for landscape irrigation by 75%, how many points can you get for Water Efficient Landscaping?
    a. 2 points.
    b. 1 point.
    c. 0 points.
    d. None of above.

35. Choose the official USGBC website(s): (Choose 2)
    a. www.USGBC.com
    b. www.LEED.org
    c. www.USGBC.org
    d. www.Greenbuild365.org

36. If you reuse, refurbish or salvage 10% (based on value) of the building materials for a NC project, how many point(s) can you get for Material Reuse?
    a. 0 points.
    b. 1 point.
    c. 2 points.
    d. 3 points.

37. For MRc4, recycled content is defined by the:
    a. USGBC
    b. ISO
    c. ICC
    d. ICBO

38. For MRc4, recycled content is based on:
    a. Cost
    b. Volume
    c. Weight
    d. None of above

39. You can get two points for Regional Material by: (Choose 2)
    a. Using building materials processed, extracted and regionally manufactured (within 300 miles of your project site) for 20% (based on cost) of the total amount of materials used on the project.
    b. Using building materials processed, extracted and manufactured regionally (within 500 miles of your project site) for 10% (based on cost) of the total amount of materials used on the project.
    c. Using building materials processed, extracted and manufactured regionally (within 500 miles of your project site) for 20% (based on cost) of the total amount of materials used on the project.
    d. Using building materials processed, extracted and manufactured regionally (within 500 miles of your project site) for 30% (based on cost) of the total amount of materials used on the project.

40. For MRc7: Certified Wood is certified by the:
    a. USGBC
    b. FSC
    c. ICBO
    d. ISO

41. To get one point for Rapidly Renewable Materials for a NC project, you should:
    a. Use 2.5% of rapidly renewable materials based on cost.
    b. Use 5% of rapidly renewable materials based on cost.
    c. Use 7.5% of rapidly renewable materials based on cost.
    d. Use 10% of rapidly renewable materials based on cost.

42. SSc1: Site Selection requires you not to develop building, parking areas, hardscape, or roads: (Choose 2)
    a. On prime farmland defined by USDA.
    b. Virgin land with an elevation lower than five feet above the elevation of 100-year flood defined by FEMA.
    c. Within 200 feet of wetland defined by US Code of Federal Regulations 40 CFR.
    d. All of above

43. For SSc3: Brownfield Redevelopment, a brownfield can be defined by (choose 2):
    a. ASTM
    b. Local, state or federal government agencies
    c. USGBC
    d. All of above

44. To get one point for SSc4.3: Alternative Transportation, you can:
    a. Provide fuel-efficient and low-emitting vehicles for 3% of the FTE (Full-Time Equivalent) occupants.
    b. Provide fuel-efficient and low-emitting vehicles for 5% of the FTE (Full-Time Equivalent) occupants.
    c. Provide preferred parking for fuel-efficient and low-emitting vehicles.
    d. Both a and c.

45. For Fundamental Refrigerant Management, you need to:
    a. Reduce the use of CFC-based refrigerants in new base building HVAC& R systems by 25%.
    b. Reduce the use of CFC-based refrigerants in new base building HVAC&R systems by 50%.
    c. Eliminate the use of CFC-based refrigerants in new base building HVAC&R systems.
    d. None of above.

46. For Enhanced Commissioning on a project over 100,000 s.f., the CxA shall:
    a. Not be an employee of the design firm.
    b. Not be an employee of the contractor holding construction contracts.
    c. Be a qualified employee of the owner.
    d. All of the above.

47. For Measurement and Verification, the M and V period shall cover:
    a. At least one year of post-construction occupancy.
    b. At least two years of post-construction occupancy.
    c. At least three years of post-construction occupancy.
    d. None of above.

48. Which of the following terms have been mentioned in EAc5: Measurement and Verification: (Choose 3)
    a. International Performance Measure and Verification Protocol (IPMVP).
    b. SCAQMD.
    c. ECMs.
    d. M&V.

49. If you increase the outdoor air ventilation rate by 20% above the rate required by ASHRAE, the following(s) will improve:
    a. Heat island effect.
    b. Thermal comfort.
    c. Indoor air quality.
    d. Chance of satisfying EAp1: Fundamental Commissioning of the Building Systems.

50. You have to delete a horizontal window louver to save money for your project. Which of the following credits may be affected: (Choose 3)
    a. EAc1: Optimize Energy Performance.
    b. EAc6: Green Power.
    c. EAc3: Enhanced Commissioning.
    d. IEQc8.2: Daylight and Views.

51. Which of the following standards has a VOC limit for architectural paints, coatings and primers applied for interior walls and ceilings?
    a. Bay Area Air Quality Management District Regulations 8.
    b. Green Seal Standard GS-11, Paints, First Edition.
    c. South Coast Air Quality Management District (SCAQMD) Rule 1113.
    d. International Performance Measure and Verification Protocol (IPMVP).

52. If your building meets the minimum requirements of Section 4 through 7 of ASHRAE 62.1-2007, Ventilation for Acceptable Indoor Air Quality (with errata but without addenda), you can get:
    a. 0 points for your project.
    b. 1 point for your project.
    c. 2 points for your project.
    d. 3 points for your project.

53. For mechanically ventilated space, to get one point for IEQc2: Increased Ventilation, you can:
    a. Increase the breathing zone outdoor air ventilation rates to all occupied spaces by at least 20% above the minimum rate required.
    b. Increase the breathing zone outdoor air ventilation rates to all occupied spaces by at least 30% above the minimum rate required.
    c. Increase the breathing zone outdoor air ventilation rates to all occupied spaces by at least 40% above the minimum rate required.
    d. Increase the breathing zone outdoor air ventilation rates to all occupied spaces by at least 50% above the minimum rate required.

54. Which of the following factors will affect Thermal Comfort?
    a. Air temperature, air speed, and relative humidity.
    b. Air temperature, air speed, floor to ceiling height, and relative humidity.
    c. Air temperature, radiant temperature, air speed, and relative humidity.
    d. None of above.

55. Which of the following will gain points for a project for both EAc1: Optimize Energy Performance and EAc2: On-site Renewable Energy?
    a. Install horizontal window louvers.
    b. Use a reflective, light colored roof.
    c. Use better wall and roof insulation.
    d. Install active solar thermal systems using collection panels.

56. Your project involves renovation of a commercial building with seven floors each of equal area. You are doing renovation work on three floors and core and shell. Your scope of work includes mechanical and plumbing systems, tenant improvements and exterior window replacement. Which of the following LEED rating systems shall you use?
    a. LEED-EB.
    b. LEED-EB and LEED-CS.
    c. LEED-CI and LEED-CS.
    d. LEED-EB and LEED-CS.

57. When should you start to prepare documentation for LEED credit submittals?
    a. In the pre-design stage.
    b. In the schematic design stage.
    c. At the beginning of the design development stage.
    d. Before construction.

58. What is the intent of EQc6.1: Controllability of Systems: Lighting?
    a. Save electricity.
    b. Occupants can customize their surroundings.
    c. Energy efficiency.
    d. To exceed ASHRAE requirements.

59. The intent of EQc5: Indoor Chemical and Pollutant Source Control is:
    a. To avoid a chemical spill.
    b. To avoid building occupants' exposure to chemical pollutants and hazardous particles.
    c. Both a and b.
    d. None of above.

60. You are renovating a 20,000 s.f. building, and adding 60,000 s.f. of new steel-frame construction. You are using reclaimed wood from a local company and replacing single pane glazing with insulated glazing manufactured by a local firm. You may be able to get credit for the following categories (Choose 3):
    a. MRc3, Materials Reuse.
    b. MRc5, Regional Materials.
    c. MRc2, Construction Waste Management.
    d. MRc7, Certified Wood.

61. Which two of the following statements are false?
    a. Your building cannot get any credits at the design phase review.
    b. You can only file an appeal after construction phase review.
    c. Your building can get LEED certification after the design phase review.
    d. If you do a design phase submittal and the project conditions change, you must submit additional information regarding the changes in the construction phase review.

62. Your LEED-registered building has complicated issues and does not fall into any established credit categories or credit submittal templates, what should you do?
    a. Contact the USGBC team directly and try to resolve the issue.
    b. Pick a similar LEED credit category and submit documentation in this category.
    c. Provide a narrative explaining your complicated issues and describe how the credit intent is met.
    d. Follow the alternate compliance path option on the LEED submittal template and describe how your building complies.

63. Which of the following should not be in your specifications to the contractors for MRc2?
    a. The amount of waste leaving the site.
    b. A description of the waste materials.
    c. The requirements of a site logistic plan.
    d. The requirement of identifying the recyclers and haulers.

64. You design a residential condominium with 300 occupants; you have included 30 bicycle racks inside the parking structure of the building. What must you do to get one point for Alternative Transportation?
    a. Increase the number of bicycle racks to 45.
    b. Place the bicycle racks inside a lockable, enclosed space.
    c. Make sure the bicycle racks are placed within 200 yards of the building.
    d. Add one shower and changing facility for each sex.

65. To get one point for MRc1.1: what surface areas do you need? (Choose 2)
    a. The landscaped areas to remain.
    b. Roof and floor areas.
    c. Window assemblies.
    d. Exterior framing and skin.

66. To get one point for SSc7.1: Heat Island Effect: Non-Roof, you must:
    a. Place 50% of the parking spaces under cover.
    b. Use paving materials with an SRI of 50 or higher.
    c. Provide an open grid pavement system for 50% of the parking surface.
    d. None of the above.

67. You are designing a roof with a slope of 1:12. To get one point for SSc7.2: Heat Island Effect: Roof, you must:
    a. Use roofing material with a SRI of at least 29 for 75% of the roof surface.
    b. Use roofing material with a SRI of at least 78 for 75% of the roof surface.
    c. Provide a vegetated roof for at least 50% of the roof surface.
    d. Either b or c.

68. For SSc8: Light Pollution Reduction, select the incorrect statement(s) from below: (Choose 2)
    a. The zones of your projects are defined by IESNA RP-33.
    b. The zones of your projects are defined by ASHRAE.
    c. Both IESNA RP-33 and ASHRAE are mentioned in this credit.
    d. LZ4 is the dark zone designation (Park and Rural Settings).

69. The imperviousness for your project is 65%. To get one point for SSc6.1: Storm Water Design, you must:
    a. Apply a storm water management to reduce 25% of the volume of storm water runoff from the two-year 24-hour design storm.
    b. Apply a storm water management to reduce 50% of the volume of storm water runoff from the two-year 24-hour design storm.
    c. Apply a storm water management to reduce 30% of the volume of storm water runoff from the two-year, 24-hour design storm.
    d. None of the above.

70. Your project is located on a previously developed site. To get one point for SSc5.1: Site Development, you need to:
    a. Limit site disturbances to 40 feet beyond the building perimeter.
    b. Restore or protect at least 50% of the site area with adapted or native vegetation.
    c. Both a and b.
    d. None of the above.

71. Which of the following is not mentioned in IEQc3.1, Construction IAQ Management Plan?
    a. Sheet Metal and Air Conditioning Contractor Association (SMACNA)
    b. Minimum Efficiency Reporting Value (MERV)
    c. ASHRAE
    d. SCAQMD

72. You can satisfy IEQp2, Environmental Tobacco Smoke (ETS) Control by:
    a. Banning smoking in the building.
    b. Placing all outdoor designated smoking area a minimum of 25' from operable windows, outdoor air intakes and entries.
    c. Both a and b.
    d. None of the above.

73. To get one point for IEQc8.1, Daylight and Views, you can:
    a. Provide a daylight illumination level of at least 25 footcandles for at least 75% of the regularly occupied areas.
    b. Provide a daylight illumination level of at least 15 footcandles for at least 90% of the regularly occupied areas.
    c. Provide a daylight illumination level of at least 20 footcandles for at least 75% of the regularly occupied areas.
    d. None of the above.

74. To get one point for IEQc7.2, Thermal Comfort:
    a. You should conduct a thermal comfort survey of building occupants within 6 months after occupants.
    b. You should conduct a thermal comfort survey of building occupants within 12 months after occupants.
    c. You should conduct a thermal comfort survey of building occupants within 18 months after occupants.
    d. You should conduct a thermal comfort survey of building occupants within a period of 6 to 18 months after occupants.

75. The following terms are mentioned in IEQc4.4, Low-Emitting Materials: (Choose 3)
    a. Urea-formaldehyde resins.
    b. Medium Density Fiberboard (MDF).
    c. SCAQMD.
    d. Fit-out, Furniture, and Equipment (FF&E).

76. For IEQc3.2, Construction IAQ Management Plan, flush-out requires:
    a. Supplying 12,000 cu.ft. of outdoor air while maintaining the humidity less than 60% and an internal temperature of 60 degrees minimum.
    b. Supplying 12,000 cu.ft. of outdoor air while maintaining the humidity less than 80% and an internal temperature of 60 degrees minimum.
    c. Supplying 14,000 cu.ft. of outdoor air while maintaining the humidity less than 60% and an internal temperature of 60 degrees minimum.
    d. Supplying 15,000 cu.ft. of outdoor air while maintaining the humidity less than 60% and an internal temperature of 60 degrees minimum.

77. You are working on a new construction project and you provide an accessible area for storage and collection of recyclables. You will:
    a. get 0 points for the Materials and Resources LEED credit.
    b. get 1 point for the Materials and Resources LEED credit.
    c. get 2 points for the Materials and Resources LEED credit.
    d. None of the above.

78. Which of the following is false?
    a. Using a special kind of window to prevent birds from flying into the windows and getting killed may add one point to your project.
    b. You can get two extra ID points for WEc3, Water Use Reduction by saving 40% water.
    c. You need to have at least one LEED AP on your project team before project registration to get one point for IDc2, LEED Accredited Professional.
    d. No matter how many LEED APs your project team has, your project can get only one point for IDc2, LEED Accredited Professional.

79. Which of the following is not included in the commissioning report?
    a. Compliance with the design intent.
    b. Documentation for O&M.
    c. Specifications.
    d. Electrical riser diagram.

80. N.R.D.C. suggests that the US cease the production of CFCs in which year?
    a. 1985
    b. 1991
    c. 1993
    d. 1995

81. Which three of the following statements are correct? (Choose 3)
    a. A LEED project team leader is in charge of the LEED project check list.
    b. A LEED project team leader sets up LEED Online.
    c. A LEED project administrator has to be a LEED AP.
    d. A LEED project team assigns responsibilities to team members, and s/he is the only one who can input information to LEED Online
    e. A LEED project administrator is the point of contact between the LEED project team and the GBCI.

82. Which statements below are incorrect? (Choose 3)
    a. Only LEED project team members can search for existing CIRs in the GBCI/USGBC database.
    b. Only USGBC members can search for existing CIRs in the GBCI/USGBC database.
    c. The public can search for existing CIRs in the GBCI/USGBC database.
    d. The LEED project team members have to call the USGBC/GBCI for existing CIRs.

83. For LEED submittal phases: (Choose 3)
    a. You can only file appeals for a construction phase review.
    b. You can file appeals for design or construction phase reviews.
    c. The LEED project team decides which package to submit at the design phase.
    d. All WE credits can be submitted at the design phase.

84. Which one of the following is NOT addressed by SS of the LEED system?
    a. Noise pollution
    b. Air pollution
    c. Water pollution
    d. CGP

85. Stabilization techniques include: (Choose 3)
    a. Mulching.
    b. Silt fencing.
    c. Temporary seeding.
    d. Permanent seeding.

86. Which of the following are required submittals for SSp1?
    a. A sedimentation and erosion control drawing
    b. A narrative detailing BMP, ESC and the responsible parties
    c. Confirmation showing compliance with local Erosion Control Standards or NPDES.
    d. Reports or inspection logs, date-stamped photos, etc.
    e. All of the above

87. For SS in the LEED system, wetlands are defined by
    a. FEMA
    b. USDA
    c. U.S. CFR
    d. Wetland Protection Act

88. For SS, basic services include, but are not limited to: (Choose 4)
    a. Banks
    b. Places of Worship
    c. Convenience Stores
    d. Bus stops
    e. Day Cares

89. Brownfields are defined by:
    a. FEMA
    b. USDA
    c. U.S. CFR
    d. Wetland Protection Act
    e. EPA

90. Zero Emission Vehicles (ZEV) are defined by the:
    a. American Council for an Energy Efficient Economy (ACEEE)
    b. U.S. CFR
    c. EPA
    d. California Air Resources Board

91. Your project is at a previously developed site, and includes 55% landscaped area planted with adapted or native vegetation, a retention pond, a wetland and a watercourse. The project's design strategies may contribute to: (Choose 4)
    a. WEc1: Water Efficient Landscaping
    b. WEc2: Innovative Wastewater Technologies
    c. WEc3: Water Use Reduction
    d. SSc5.1: Site Development: Protect or Restore Habitat
    e. SSc6.1: Storm Water Design: Quantity Control
    f. SSc7.1: Heat Island Effect: Non-Roof

92. Natural and mechanical ventilation in a building may contribute to: (Choose 3)
    a. IEQc2: Increased Ventilation
    b. IEQp1: Minimum IAQ Performance
    c. EAp1: Fundamental Commissioning of the Building Energy Systems
    d. SSc1: Site Selection
    e. All of the above

93. Increasing graywater use, increasing rainwater harvesting and decreasing the demand on local water aquifers may contribute to:
    a. SSc6, Storm Water Design
    b. WEc1, Water Efficient Landscaping
    c. WEc2, Innovative Wastewater Technologies
    d. WEc3, Water Use Reduction
    e. WEc4, Process Water Use Reduction
    f. All of the above

94. Which of the following statements are correct? (Choose 2)
    a. Water use for a conventional water closet is 1.6 gpf.
    b. The water use baseline for a residential lavatory is 2.5 gpm at 60 psi.
    c. The water use baseline for a public lavatory is 0.5 gpm at 60 psi.
    d. All of the above

95. Which of the following statements regarding water use for a water closet are correct? (Choose 2)
    a. 1.6 gpf for baseline
    b. 1.0 gpf for low-flow
    c. 1.1 gpf for HET dual flush (low-flush)
    d. 0.5 gpf for HET, foam flush
    e. All of the above

96. Which of the following statements are correct? (Choose 3)
    a. WaterSense is developed per EPA and ASHRAE standards.
    b. WaterSense is a partnership program sponsored by the EPA.
    c. WaterSense Standards exceed IPC and UPC requirements in some cases.
    d. Use of WaterSense fixtures may qualify your project for LEED credits.

97. The Energy Policy Act of 1992 is related to:
    a. WE
    b. SS
    c. EA
    d. IEQ
    e. ID

98. Which three of the following statements are correct?
    a. Showers are considered to be a 480 second duration for residential projects, and a 280 second duration for non-residential projects.
    b. High-efficiency (HE) fixtures include: single gravity, high-efficiency toilets (HET) and high-efficiency urinals (HEU).
    c. The flow rate for ultra low-flow lavatory use is 0.5 gpm.
    d. The flow rate for showers is 2.5 gpm at 60 psi per shower stall.
    e. The flow rate for a kitchen sink is 2.2 gpm at 60 psi.

99. A project has 100 tons of demolition waste. 30 tons will be donated, 20 tons will be re-used on-site, and 25 tons will be recycled. Which two of the following statements are correct?
    a. The project can get one point for MRc2, Construction Waste Management.
    b. The 30 tons of recycled material can be included in MRc4, Recycled Content.
    c. Since the donated materials cannot be included in the diverted waste, this project cannot gain one point for MRc2, Construction Waste Management.
    d. You cannot count the 20 tons of re-used materials toward MRc3, Materials Reuse if you already included the material in MR Credit 2, Construction Waste Management.

100. MEP equipment can be included in:
    a. MRc1: Building Reuse
    b. MRc2: Construction Waste Management
    c. MRc3: Materials Reuse
    d. MRc4: Recycled Content

IV. Answers for the LEED Green Associate Mock Exam

If you answer 60 of the 100 questions correctly, you have passed the mock exam. Based on our readers' feedback, this set of mock exam is HARDER than the real LEED Green Associate Exam.

We want it to be harder to make the readers nervous and force them to go back and review the study materials in my book a few more times.

This approach is apparently working. An overwhelming majority of my readers pass the LEED Green Associate Exam on the first try. It is better safe than sorry. A set of hard mock exam can be very helpful: What does not kill you makes you stronger.

We want your LEED knowledge to peak on the day of the LEED Green Associate Exam, NOT before or after. One way to achieve this goal is to make you nervous and keep the pressure on, and you will be willing to go back and seriously study and keep reviewing the materials in my book.

We have published 2 additional sets of mock exams as a separate book, *LEED GA Mock Exams*. To balance your exam prep effort, the sample tests in "LEED GA Mock Exams" are very close to the real exams.

1. Answer: d. See LEED checklist. See appendix for link to download the checklists from USGBC website.
2. Answer: b
3. Answer: d
4. Answer: c
5. Answer: a
6. Answer: b
7. Answer: d
8. Answer: a
9. Answer: c
10. Answer: d
11. Answer: d
12. Answer: d
13. Answer: c and d
14. Answer: a
15. Answer: b
16. Answer: b
17. Answer: d
18. Answer: a
19. Answer: a. A job captain is a very junior member of a project team; you cannot get any LEED point if the LEED AP is not a principal of your team.
20. Answer: c
21. Answer: a and b
22. Answer: d
23. Answer: d
24. Answer: c
25. Answer: a
26. Answer: b
27. Answer: a

28. Answer: b
29. Answer: b
30. Answer: d
31. Answer: b
32. Answer: a
33. Answer: c
34. Answer: a
35. Answer: c and d
36. Answer: d. See MRc3, Materials Reuse
37. Answer: b
38. Answer: a
39. Answer: a and c
40. Answer: b
41. Answer: a
42. Answer: a and b
43. Answer: a and b
44. Answer: d
45. Answer: c
46. Answer: d
47. Answer: a
48. Answer: a, c and d
49. Answer: c
50. Answer: a, b and d
51. Answer: b. See IEQc4.2, Low Emitting Materials
52. Answer: a. See IEQp1, Minimum IAQ Performance
53. Answer: b
54. Answer: c
55. Answer: d
56. Answer: c
57. Answer: a
58. Answer: b
59. Answer: b
60. Answer: a, b and c
61. Answer: b and c
62. Answer: d
63. Answer: c
64. Answer: a. See SS Credit 4.2. For residential buildings, you need 15% bicycle racks.
65. Answer: b and d
66. Answer: a
67. Answer: d
68. Answer: b and d
69. Answer: a
70. Answer: b
71. Answer: d
72. Answer: c
73. Answer: a
74. Answer: d
75. Answer: a, b and d
76. Answer: c

77. Answer: a
78. Answer: b
79. Answer: d
80. Answer: d
81. Answer: a, b, and e
82. Answer: a, c, and d
83. Answer: b, c, and d
84. Answer: a
85. Answer: a, c, and d
86. Answer: e
87. Answer: c
88. Answer: a, b, c, and e
89. Answer: e
90. Answer: d
91. Answer: a, d, e, f. WEc2, Innovative Wastewater Technologies and WEc3, Water Use Reduction are for INTERIOR water.
92. Answer: a, b, and c
93. Answer: f
94. Answer: a and c
95. Answer: a and c
96. Answer: b, c, and d
97. Answer: a
98. Answer: b, c, and e
99. Answer: a and d. b is incorrect because "recycled" in this question refers to "waste recycled" rather than the "recycled content of materials used."
100. Answer: b

V. How were the LEED Green Associate mock exams created?

The actual LEED Green Associate Exam has 100 questions and you must finish it within two hours. The raw exam score is converted to a scaled score ranging from 125 to 200. The passing score is 170 or higher.

I tried to be scientific when selecting the mock exam questions, so I based the number of questions for each credit category roughly on the number of points that you can get for that category. The level difficulty for each question was designed to match the 12 sample questions that can be downloaded from the official GBCI website. Feedback from our readers has indicated that this mock exam is relatively easy when compared to the actual LEED Green Associate Exam .

VI. Where can I find the latest official sample questions for the LEED Green Associate exam?

Answer: You can find them, as well as the exam content from the candidate handbook, at:
http://www.gbci.org/main-nav/professional-credentials/resources/candidate-handbooks.aspx

VII. LEED Green Associate Exam registration

1. **How to register for the LEED Green Associate Exam?**
   Answer: Per the GBCI, you must create an Eligibility ID at www.GBCI.org. Select the "Schedule an Exam" menu to set up an exam time and date with Prometric. You can reschedule or cancel the LEED Green Associate Exam at www.prometric.com/gbci with your Prometric-issued confirmation number for the exam. You need to bring two forms of ID to the exam site. See www.prometric/gbci for a list of exam sites. Call 1-800-795-1747 (within the US) or 202-742-3792 (Outside of the US) or e-mail exam@gbci.org if you have any questions.

2. **Important Note:** You can download the "LEED Green Associate Candidate Handbook" from the GBCI website and get all the latest details and procedures. Ideally you should download it and read it carefully at least three weeks before your exam. See link below:
   http://www.gbci.org/main-nav/professional-credentials/resources/candidate-handbooks.aspx

Chapter 6
Frequently Asked Questions (FAQ) and Other Useful Resources

The following are tips on how to pass the LEED exam on the first try and in one week. I also include my responses to some readers' questions. They may help you:

1. I found the reference guide way too tedious. Can I only read your book and just refer to the USGBC reference guide (if one is available for the exam I am taking) when needed?

Response: Yes. That is one way to study.

2. Is one week really enough for me to prepare for the exam while I am working?

Response: Yes, if you can put in 40 to 60 hours during the week, study hard and you can pass the exam. This exam is similar to a history or political science exam; you need to MEMORIZE the information. If you take too long, you will probably forget the information by the time you take the test.

In my book, I give you tips on how to MEMORIZE the information, and I have already highlighted/underlined the most important information that you definitely have to MEMORIZE to pass the exam. It is my goal to use this book to help you to pass the LEED exam with the minimum time and effort. I want to make your life easier.

3. Would you say that if I buy your LEED Exam Guide Series books, I could pass the exam using no other study materials? The books sold on the USGBC website run in the hundreds of dollars, so I would be quite happy if I could buy your book and just use that.

Response: First of all, there are readers who have passed the LEED Exam by reading only my books in the LEED Exam Guides Series (www.ArchiteG.com). My goal is to write one book for each of the LEED exams, and make each of my books stand alone to prepare people for one specific LEED exam.

Secondly, people learn in many different ways. That is why I have added some new advice below for people who learn better by doing practice tests.

If you do the following things, you have a very good chance of passing the LEED exam (NOT a guarantee, nobody can guarantee you will pass):

a. If you study, understand and MEMORIZE all of the information in my book, and do NOT panic when you run into problems you are not familiar with, and use the guess strategy in my book, then you have a very good chance of passing the exam.

You need to UNDERSTAND and MEMORIZE the information in the book and score almost a perfect score on the mock exam in this book. This book will give you the BULK of the most CURRENT information that you need for the specific LEED exam you are taking. You HAVE to know the information in my book in order to pass the exam.

b. If you have not done any LEED projects before, I suggest you also go to the USGBC website and download the latest LEED credit templates for the LEED rating system related to the LEED exam you are taking. Read the templates and become familiar with them. This is important. See link below:
http://www.usgbc.org/DisplayPage.aspx?CMSPageID=222

c. If you want to be safe and do additional sample tests to find out if you are ready for the real exam, we have other books on various LEED mock exams, including *LEED GA Mock Exams*. Check them out at:
www.**GreenExamEducation**.com

The LEED GA mock exam set in this book is HARDER than the real LEED Green Associate Exam. To balance your exam prep effort, the sample tests in *LEED GA Mock Exams* are very close to the real exams.

The LEED exam is NOT an easy exam, but anyone with a 7th grade education should be able to study and pass the LEED exam if he prepares correctly.

If you have extra time and money, the other books I would recommend are *LEED GA Mock Exams* and the USGBC reference guide, the official book for the LEED NC exam. I know some people who did not even read the reference guide from cover to cover when they took the exam. They just studied the information in my book, and only referred to the reference guide to look up a few things, and they passed on the first try. Some of my readers have even passed WITHOUT reading the USGBC reference guide AT ALL.

4. I am preparing for the LEED exam. Do I need to read the 2" thick reference?

Response: See answer above.

5. For LEED v3.0, will the total number of points be more than 110 in total if a project gets all of the extra credits and all of the standard credits?

Response: No. For LEED v3.0, there are <u>100</u> base points and <u>10</u> possible bonus points. There are many ways to get bonus points (extra credits or exemplary performance), but you can have a maximum number of <u>6 ID</u> bonus points and <u>4 Regional Priority</u> bonus points. So, the maximum points for ANY project will be <u>110</u>.

On another note, the older versions of LEED rating systems all have less than 110 possible points except LEED for **Home**, which has 136 possible points.

6. For the exam, do I need to know the project phase in which a specific prerequisite/credit takes place? (i.e., pre-design, schematic design, etc.)

Response: The information on the project phase (NOT LEED submittal phase) for each prerequisite/credit is NOT mentioned in the USGBC reference guide, but it is covered in the USGBC workshops. If it is important enough for the USGBC workshops to cover, then it may show up on the actual LEED exam.

Most, if not all, other third party books completely miss this important information. I cover it for each

prerequisite/credit in my book for the LEED exam because I think it is very important.
Some people THINK that the LEED exam ONLY tests information covered by the USGBC reference guide. They are wrong.

The LEED exam does test information NOT covered by the USGBC reference guide at all. This may include the process of LEED submittal and project team coordination, etc.

I would MEMORIZE this information if I were you, because it may show up on the LEED exam. Besides, this information is not hard to memorize once you understand it, and you need to know it to do actual LEED submittal work anyway.

7. Are you writing new versions of books for the new LEED exams? What new books are you writing?

Response: Yes, I am working on other books in the LEED Exam Guide Series. I will be writing one book for each of the LEED exam. See LEEDSeries.com for more information.

8. Important documents that you need to download for <u>free</u>, become familiar with and <u>memorize</u>:

Note: GBCI and USGBC change the links to their document every now and then, so, by the time you read this book, they may have changed some of the following links. You can simply go to their main website, search for the document with its name, and should be able to find the most current link. You can use the same technique to search for documents by other organizations.

The main website for the GBCI is:
http://www.gbci.org/

The main website for the USGBC is:
http://www.usgbc.org/

a. Every LEED exam **always tests** Credit Interpretation Request (CIR). Download the guidelines for CIR customers, read and <u>memorize</u> it:
http://www.gbci.org/Certification/Resources/cirs.aspx

b. Every LEED exam **always tests** project team coordination. Download *Sustainable Building Technical Manual: Part II,* by Anthony Bernheim and William Reed (1996), read and <u>memorize</u> it:
http://www.gbci.org/Files/References/Sustainable-Building-Technical-Manual-Part-II.pdf

c. Project registration application and LEED certification process:
http://www.usgbc.org/DisplayPage.aspx?CMSPageID=1497

d. LEED Online:
http://www.usgbc.org/DisplayPage.aspx?CMSPageID=277
https://www.gbci.org/DisplayPage.aspx?CMSPageID=137

9. Important documents that you need to download for <u>free</u>, and become <u>familiar</u> with:

a. *LEED for Operations and Maintenance Reference Guide-Introduction* (U.S. Green Building Council, 2008)

https://www.usgbc.org/ShowFile.aspx?DocumentID=4512

b. *LEED for Operations and Maintenance Reference Guide-Glossary* (U.S. Green Building Council, 2008) http://www.gbci.org/Files/References/LEED-for-Operations-and-Maintenance-Reference-Guide-Glossary.pdf

c. *LEED for Homes Rating System* (U.S. Green Building Council, 2008) http://www.gbci.org/Files/References/LEED-for-Homes-Rating-System.pdf

Pay special attention to the list of **abbreviations and acronyms** on pages 105–106 and a helpful **glossary of terms** on pages 107–114.

d. *Cost of Green Revisited,* by Davis Langdon (2007) http://www.gbci.org/Files/References/Cost-of-Green-Revisited.pdf

e. *The Treatment by LEED® of the Environmental Impact of HVAC Refrigerants (*LEED Technical and Scientific Advisory Committee, 2004) http://www.gbci.org/Files/References/The-Treatment-by-LEED-of-the-Environmental-Impact-of-HVAC-Refrigerants.pdf

f. *Guidance on Innovation and Design (ID) Credits* (US Green Building Council, 2004) http://www.gbci.org/Files/References/Guidance-on-Innovation-and-Design-Credits.pdf

10. Do I need to take many practice questions to prepare for a LEED exam?

Response: There is NO absolutely correct answer to this question. People learn in many different ways. Personally, I am NOT crazy about doing many practice questions. Consider if you do 700 practice questions, not only must you read them all, but each question has at least 4 choices. That totals to at least 2,800 choices, which is a great deal of reading. I have seen some third-party materials that have 1,200 practice questions. That will require even MORE time to go over the materials.

I prefer to spend most of my time reading, digesting, and really understanding the fundamental materials, and MEMORIZE them naturally by rereading the materials multiple times. This is because the fundamental materials for ANY exam will NOT change, and the scope of the exam will NOT change for the same main version of the test (until the exam moves to the next advanced version). However, there are many ways to ask you questions.

If you have a limited amount of time for preparation, it is more efficient for you to focus on the fundamental materials and actually <u>master</u> the knowledge that GBCI wants you to learn. If you can do that, then no matter how GBCI changes the exam format or how GBCI asks the questions, you will do fine in the exam.

Strategy 101 for the LEED Green Associate Exam is that you must recognize that you have only a limited amount of time to prepare for the exam. Therefore, you must concentrate on the most important contents of the LEED Green Associate Exam.

The key to passing the LEED Green Associate Exam, or any other exam, is to know the scope of the exam, and not to read too many books. Select one or two helpful books and focus on them. You must understand the content and memorize it. For your convenience, I have underlined the fundamental information that I think is very important. You definitely need to memorize all the information that I have underlined. You should try to understand the content first, and then memorize the content of the book by rereading it. This is a much better way than "mechanical" memory without understanding.

Most people fail the exam NOT because they are unable to answer the few "advanced" questions on the exam, but because they have read the information but can NOT recall it on the day of the exam. They spend too much time preparing for the exam, drag the preparation process on too long, seek too much information, go to too many Web sites, do too many practice questions and too many mock exams (one or two sets of mock exams are probably sufficient), and spread themselves too thin. They end up missing out on the most important information of the LEED exam, and they will fail.

To me, Memorization and Understanding work hand-in-hand. Understanding always comes first. If you really understand something, then Memorization is easy.

For example, I'll read a book's first chapter very slowly but make sure I <u>really</u> understand everything in it, no matter how long it takes. I do NOT care if others are faster readers than I. Then, I reread the first chapter again. This time, the reading is so much easier, and I can read it much faster. Then I try to retell the contents, focusing on substance, not the format or any particular order of things. This is a very good way for me to understand and digest the material, while <u>absorbing</u> and <u>memorizing</u> the content.

I then repeat the same procedure for each chapter, and then reread the book until I take the exam. This achieves two purposes:

a. I keep reinforcing the important materials that I already have memorized and fight against the human brain's natural tendency to forget things.

b. I also understand the content of the book much better by reading it multiple times.

If I were to attempt to memorize something without understanding it first, it would be very difficult for me to do so. Even if I were to memorize it, I would likely forget it quickly.

Appendixes

1. Default occupancy factors

Occupancy	Gross sf per occupant	
	Transient Occupant	**FTE**
Educational, Daycare	630	105
Educational, K–12	1,300	140
Educational, Postsecondary	2,100	150
Grocery store	550	115
Hotel	1,500	700
Laboratory or R&D	400	0
Office, Medical	225	330
Office, General	250	0
Retail, General	550	130
Retail or Service (auto, financial, etc.)	600	130
Restaurant	435	95
Warehouse, Distribution	2,500	0
Warehouse, Storage	20,000	0

Note: This table is for projects (like CS) where the final occupant count is not available. If your project's occupancy factors are not listed above, you can use a comparable building to show the average gross sf per occupant for your building's use.

2. Important resources and further study materials you can download for <u>free</u>

Energy Performance of LEED® for New Construction Buildings: Final Report, by Cathy Turner and Mark Frankel (2008):
http://www.gbci.org/Files/References/Energy-Performance-of-LEED-for-New-Construction-Buildings-Final-Report.pdf

Foundations of the Leadership in Energy and Environmental Design Environmental Rating System: A Tool for Market Transformation (LEED Steering Committee, 2006):
http://www.gbci.org/Files/References/Foundations-of-the-Leadership-in-Energy-and-Environmental-Design-Environmental-Rating-System-A-Tool-for-Market-Transformation.pdf

AIA Integrated Project Delivery: A Guide (www.aia.org):
http://www.aia.org/contractdocs/AIAS077630

Review of ANSI/ASHRAE Standard 62.1-2007: Ventilation for Acceptable Indoor Air Quality, by Brian Kareis:
http://www.workplace-hygiene.com/articles/ANSI-ASHRAE-3.html

Best Practices of ISO-14021: Self-Declared Environmental Claims, by Kun-Mo Lee and Haruo Uehara (2003):

http://www.ecodesign-company.com/
documents/BestPracticeISO14021.pdf

Bureau of Labor Statistics (www.bls.gov)

International Code Council (www.iccsafe.org)

Americans with Disabilities Act (ADA): Standards for Accessible Design (www.ada.gov):
http://www.ada.gov/stdspdf.htm

GSA 2003 Facilities Standards (General Services Administration, 2003):
http://www.gbci.org/Files/References/GSA-2003-facilities-standards.pdf

Guide to Purchasing Green Power (Environmental Protection Agency, 2004):
http://www.gbci.org/Files/References/Guide-to-Purchasing-Green-Power.pdf

USGBC Definitions:
https://www.usgbc.org/ShowFile.aspx?DocumentID=5744

3. Annotated bibliography

Chen, Gang. **LEED GA MOCK EXAMS:** *Questions, Answers, and Explanations: A Must-Have for the LEED Green Associate Exam, Green Building LEED Certification, and Sustainability*. ArchiteG, Inc, 2010. This is a companion to *LEED Green Associate Exam Guide (LEED GA)*. It includes 200 questions, answers, and explanation, and is very close to the real LEED Green Associate Exam.

4. Valuable Web sites and links

a. The Official Web sites for the U.S. Green Building Council (USGBC):
http://www.usgbc.org/
http://www.Greenbuild365.org

Pay special attention to the purpose of <u>LEED Online, LEED project registration, LEED certification content, LEED reference guide introductions, LEED rating systems, and checklists</u>.

You can download or purchase the following useful documents from the USGBC or GBCI Web site: Latest and official LEED exam candidate handbooks including an exam content outline and sample questions:
http://www.gbci.org/main-nav/professional-credentials/resources/candidate-handbooks.aspx

LEED Reference Guides, **LEED Rating System Selection Policy**, and various versions of LEED Green Building Rating Systems and Project Checklist:
http://www.usgbc.org/projecttools

Read the **LEED Rating System Selection Policy** <u>at least three times</u>, because it is VERY important, and it tells you which LEED system to use.

LEED 2009 Vision and Executive Summary:
http://www.usgbc.org/ShowFile.aspx?DocumentID=4121

USGBC issue LEED Addenda for various LEED Green Building Rating **Systems** and **reference guides** on a quarterly basis. **Make sure you download the latest LEED Addenda** related to your exam and read them at least three times. See link below for detailed information:
http://www.usgbc.org/addenda

b. Natural Resources Defense Council:
    http://www.nrdc.org/

c. Environmental Construction + Design - Green Book (Offers print magazine and online environmental products and services resources guide):
    http://www.edcmag.com/greenbook

d. Cool Roof Rating Council Web site:
    http://www.coolroofs.org

5. Important Items Covered by the Second Edition of *Green Building and LEED Core Concepts Guide*

Starting on December 1, 2011, GBCI will begin to draw LEED Green Associate Exam questions from the second edition of *Green Building and LEED Core Concepts Guide*. The following are some "new" and important items covered by this edition:

adaptive reuse: Designing and constructing a building to accommodate a future use that is different from its original use.

biomimicry: Learning from nature and designing systems using principles that have been tested in nature for millions of years.

carbon overlay: LEED credit weighting based on each credit's impact on reducing carbon footprint.

charrettes: Intensive (design) workshops.

cradle to cradle: A method where materials are used in a closed system and generate no waste.

cradle to grave: A process that examines materials from their point of extraction to disposal.

closed system: There is no "away." Everything goes somewhere within the system, the waste generated by a process becomes the "food" of another process. Nature is a closed system.

embodied energy: The total energy consumed by extracting, harvesting, manufacturing, transporting, installing, and using a material through its entire life cycle.

ENERGY STAR's Portfolio Manager: An online management tool for tracking and evaluating water and energy use. An ENERGY STAR Portfolio Manager score of 50 means a building is at national average energy use level for its category. A score higher than 50 means a building is more energy efficient than the national average energy use level for its category. The higher the score, the better.

evapotranspiration: Loss of water due to evaporation.

externalities: Benefits or costs that are NOT part of a transaction.

feedback loop: Information flows within a system that allows the system to adjust itself. A thermostat or melting snow is an example of negative feedback loop. Population growth, heat island effect, or climate change is a positive feedback loop. Positive feedback loop can create chaos in a system.

International Green Construction Code (IGCC): A national model green building code published by International Code Council (ICC).

integrated process: Emphasizes communications and interactions among stakeholders throughout the life of a project. Integrated process is a holistic decision making process based on systems thinking and life cycle approach.

interative process: A repetitive and circular process that helps a team to define goals and check ideas against these goals.

Integrated Pest Management (IPM): A sustainable approach to pest management.

LEED interpretations: Precedent-setting (project credit interpretation) rulings. A project team can opt into the LEED interpretation process when submitting an inquiry to GBCI.

leverage points: Places where a small interventions can generate big changes.

life cycle approach: Looking at a product or building through its entire life cycle.

life cycle assessment (LCA): Use life cycle thinking in environmental issues.

life cycle costing: Looking at the cost of purchasing and operating a building or product, and the relative savings.

low impact development (LID): A land development approach mimicking natural systems and managing storm water as close to the source as possible.

"Net-Zero": A project, which doesn't use any more resources than what it can produce. Similar concepts include carbon neutrality and water balance.

negative feedback loop: A signal for the system to stop changes when a response is not needed anymore.

open system: Resources are brought from the outside, consumed, and then disposed of as waste to the outside.

permaculture: Designing human habitats and agriculture systems based on models and relationships found in nature.

positive feedback loop: A stimulus causes an effect and encourages the loop to produce more of this effect.

Prius effect: Provides real time feedback of energy use so that users can adjust behaviors to save energy.

Project CIRs: LEED credit interpretation rulings for specific project circumstances.

retrocommissioning: A building tune-up that restores efficiency and improves performance.

regenerative: Regenerative buildings and communities evolve with living systems and help to renew resources and life. Regenerative projects generate electricity and sell the excess back to the grid, as well as return water to nature, which is cleaner than it was before use.

systems thinking: In a system, each component affects many other components. They are all related to each other.

Wingspread Principles on the US Response to Global Warming: A set of principles signed by organizations and individuals to express their commitment to address global warming. It calls for 60% to 80% reduction of green house gas emission by midcentury (based on 1990 levels).

Back page promotion:

LEED GA MOCK EXAMS: *Questions, Answers, and Explanations: A Must-Have for the LEED Green Associate Exam, Green Building LEED Certification, and Sustainability*, Book 8, LEED Exam Guide series, ArchiteG.com (Published August 6, 2010)

LEED O&M MOCK EXAMS: *Questions, Answers, and Explanations: A Must-Have for the LEED O&M Exam, Green Building LEED Certification, and Sustainability*, Book 9, LEED Exam Guide series, ArchiteG.com (Published October 6, 2010)

How to order these books:
You can order the books listed above at:
http://www.GreenExamEducation.com

OR
http://www.ArchiteG.com

Following are some detailed descriptions of each text:

LEED Exam Guide series
Comprehensive Study Materials, Sample Questions, Mock Exam,
Building LEED Certification, and Going Green

LEED (Leadership in Energy and Environmental Design) is the most important trend in development and is revolutionizing the construction industry. It has gained tremendous momentum and has a profound impact on our environment. From the LEED Exam Guide series, you will learn how to:

1. Pass the LEED Green Associate Exam and various other LEED AP+ exams (each book will help you with a specific LEED exam).

2. Register and certify a building for LEED certification.

3. Understand the intent of each LEED prerequisite and credit.

4. Calculate points for a LEED credit.

5. Identify the responsible party for each prerequisite and credit.
6. Earn extra credits (exemplary performance) for LEED.

7. Implement the local codes and building standards for prerequisites and credits.

8. Receive points for categories not yet clearly defined by the USGBC.

There is currently NO official GBCI book on any of the LEED exams, and most of the existing books on LEED and LEED AP+ are too expensive and too complicated to be practical or helpful. The pocket guides in the LEED Exam Guide series fill in the blanks, demystify LEED, and uncover the tips, codes, and jargon for LEED, as well as the true meaning of "going green." They will set up a solid foundation and fundamental framework of LEED for you. Each book in the LEED Exam Guide series covers every aspect of one or more specific LEED rating system in plain and concise language, and makes this information understandable to anyone.

These pocket guides are small and easy to carry to read when time permits. They are indispensable books for everyone: administrators; developers; contractors; architects; landscape architects; civil, mechanical, electrical, and plumbing engineers; interns; drafters; designers; and other design professionals.

Why is the LEED Exam Guide series needed?

A number of books are available that you can use to prepare for the LEED exams. Consider the following:

1. USGBC reference guides. You need to select the correct version of the reference guide for your exam.

The USGBC reference guides are comprehensive, but they give too much information. For example, *The LEED 2009 Reference Guide for Green Building Design and Construction (BD&C)* has approximately 700 oversized pages. Many of the calculations in the books are too detailed for the exam. The books are also expensive (approximately $200 each, so most people may not buy them for their personal use, but instead, will seek to share an office copy).

Reading a reference guide from cover to cover is good if you have the time. The problem is that very few people actually have the time to read the whole reference guide. Even if you do read the whole guide, you may not remember the important issues required to pass the LEED exam. You need to reread the material several times before you can remember much of it.

Reading a reference guide from cover to cover without a guidebook is a difficult and inefficient way of preparing for the LEED exams, because you do NOT know what USGBC and GBCI are looking for in the exam.

2. The USGBC workshops and related handouts are concise, but they do not cover extra credits (exemplary performance). The workshops are expensive, costing approximately $450 each.

3. Various books published by third parties are available on Amazon. However, most of them are not very helpful.

There are many books on LEED, but not all are useful.

Each book in the LEED Exam Guide series will fill in the blanks and become a valuable, reliable source.

a. They will give you more information for your money. Each of the books in the LEED Exam Guide series provides more information than the related USGBC workshops.

b. They are exam-oriented and more effective than the USGBC reference guides.

c. They are better than most, if not all, of the other third-party books. They give you comprehensive study materials, sample questions and answers, mock exams and answers, and critical information on building LEED certification and going green. Other third-party books only provide a fraction of this information.

d. They are comprehensive yet concise, small, and easy to carry around. You can read them whenever you have a few spare minutes.

e. They are great timesavers. I have highlighted the important information that you need to understand and MEMORIZE. I also make some acronyms and short sentences to help you easily remember the credit names.

You should devote about 1 to 2 weeks of full-time study to pass each of the LEED exams. I have met people who have spent only 40 hours of study time and passed the exams.

You can find sample texts and other information about the LEED Exam Guide series listed under the Amazon customer discussion section for each available book.

What others are saying about *LEED GA Mock Exams Guide* (Book 8, LEED Exam Guide series):

"Great news, I passed!!! As an educator and business professional I would absolutely recommend this book to anyone looking to take and pass the LEED Green Associate exam on the first attempt."
—Luke Ferland

"Elite runners will examine a course, running it before they race it...This book is designed to concentrate on increasing the intensity of your study efforts, examine the course, and run it before you race it..."
—Howard Patrick (Pat) Barry, AIA NCARB

"Like many similar test prep guides, Mr. Chen cites the resources that will be useful to study. But he goes beyond this and differentiates which ones must memorize and those you must be at least familiar with. "
—NPacella

"Read *LEED GA Mock Exams* before you start studying other resource materials. It will serve to bring your attention to the information that you are most likely to be asked on the exam as you come across it in your studying. "
—Mike Kwon

"I found these exams to be quite tougher compared to the others I took a look at, which is good as it made me prepare for the worst I would definitely recommend using these mock exams. I ultimately passed with 181... "
—Swankysenor

Building Construction
Project Management, Construction Administration, Drawings, Specs, Detailing Tips, Schedules, Checklists, and Secrets Others Don't Tell You (Architectural Practice Simplified, 2nd edition)

Learn the Tips, Become One of Those Who Know Building Construction and Architectural Practice, and Thrive!

For architectural practice and building design and construction industry, there are two kinds of people: those who know, and those who don't. The tips of building design and construction and project management have been undercover—until now.

Most of the existing books on building construction and architectural practice are too expensive, too complicated, and too long to be practical and helpful. This book simplifies the process to make it easier to understand and uncovers the tips of building design and construction and project management. It sets up a solid foundation and fundamental framework for this field. It covers every aspect of building construction and architectural practice in plain and concise language and introduces it to all people. Through practical case studies, it demonstrates the efficient and proper ways to handle various issues and problems in architectural practice and building design and construction industry.

It is for ordinary people and aspiring young architects as well as seasoned professionals in the construction industry. For ordinary people, it uncovers the tips of building construction; for aspiring architects, it works as a construction industry survival guide and a guidebook to shorten the process in mastering architectural practice and climbing up the professional ladder; for seasoned architects, it has many checklists to refresh their memory. It is an indispensable reference book for ordinary people, architectural students, interns, drafters, designers, seasoned architects, engineers, construction administrators, superintendents, construction managers, contractors, and developers.

You will learn:
1. How to develop your business and work with your client.
2. The entire process of building design and construction, including programming, entitlement, schematic design, design development, construction documents, bidding, and construction administration.
3. How to coordinate with governing agencies, including a county's health department and a city's planning, building, fire, public works departments, etc.
4. How to coordinate with your consultants, including soils, civil, structural, electrical, mechanical, plumbing engineers, landscape architects, etc.
5. How to create and use your own checklists to do quality control of your construction documents.
6. How to use various logs (i.e., RFI log, submittal log, field visit log, etc.) and lists (contact list, document control list, distribution list, etc.) to organize and simplify your work.
7. How to respond to RFI, issue CCDs, review change orders, submittals, etc.
8. How to make your architectural practice a profitable and successful business.

Planting Design Illustrated
A Must-Have for Landscape Architecture: A Holistic Garden Design Guide with Architectural and Horticultural Insight, and Ideas from Famous Gardens in Major Civilizations

One of the most significant books on landscaping!

This is one of the most comprehensive books on planting design. It fills in the blanks of the field and introduces poetry, painting, and symbolism into planting design. It covers in detail the two major systems of planting design: formal planting design and naturalistic planting design. It has numerous line drawings and photos to illustrate the planting design concepts and principles. Through in-depth discussions of historical precedents and practical case studies, it uncovers the fundamental design principles and concepts, as well as the underpinning philosophy for planting design. It is an indispensable reference book for landscape architecture students, designers, architects, urban planners, and ordinary garden lovers.

What Others Are Saying About *Planting Design Illustrated* …

"I found this book to be absolutely fascinating. You will need to concentrate while reading it, but the effort will be well worth your time."
—Bobbie Schwartz, former president of APLD (Association of Professional Landscape Designers) and author of *The Design Puzzle: Putting the Pieces Together.*

"This is a book that you have to read, and it is more than well worth your time. Gang Chen takes you well beyond what you will learn in other books about basic principles like color, texture, and mass."
—Jane Berger, editor & publisher of gardendesignonline

"As a longtime consumer of gardening books, I am impressed with Gang Chen's inclusion of new information on planting design theory for Chinese and Japanese gardens. Many gardening books discuss the beauty of Japanese gardens, and a few discuss the unique charms of Chinese gardens, but this one explains how Japanese and Chinese history, as well as geography and artistic traditions, bear on the development of each country's style. The material on traditional Western garden planting is thorough and inspiring, too. *Planting Design Illustrated* definitely rewards repeated reading and study. Any garden designer will read it with profit."
—Jan Whitner, editor of the *Washington Park Arboretum Bulletin*

"Enhanced with an annotated bibliography and informative appendices, *Planting Design Illustrated* offers an especially "reader friendly" and practical guide that makes it a very strongly recommended addition to personal, professional, academic, and community library gardening & landscaping reference collection and supplemental reading list."
—Midwest Book Review

"Where to start? *Planting Design Illustrated* is, above all, fascinating and refreshing! Not something the lay reader encounters every day, the book presents an unlikely topic in an easily digestible, easy-to-follow way. It is superbly organized with a comprehensive table of contents, bibliography, and appendices. The writing, though expertly informative, maintains its accessibility throughout and is a joy to read. The detailed and beautiful illustrations expanding on the concepts presented were my favorite portion. One of the finest books I've encountered in this contest in the past 5 years."
—Writer's Digest 16th Annual International Self-Published Book Awards Judge's Commentary

"The work in my view has incredible application to planting design generally and a system approach to what is a very difficult subject to teach, at least in my experience. Also featured is a very beautiful philosophy of garden design principles bordering poetry. It's my strong conviction that this work needs to see the light of day by being published for the use of professionals, students & garden enthusiasts."
—Donald C. Brinkerhoff, FASLA, chairman and CEO of Lifescapes International, Inc.

Index